Diagnosing Taste and Odor Problems

Source Water and Treatment Field Guide

Edited by Gary A. Burlingame

Authors: Stephen D.J. Booth, Auguste Bruchet, Andrea M. Dietrich, Daniel L. Gallagher, Djanette Khiari, I.H. (Mel) Suffet, and Sue B. Watson

American Water Works Association

Diagnosing Taste and Odor Problems—Source Water and Treatment Field Guide
Copyright © 2011 American Water Works Association

AWWA Publications Manager: Gay Porter De Nileon
Senior Technical Editor: Melissa Valentine
Production Editor: Cheryl Armstrong

Disclaimer

Library of Congress Cataloging-in-Publication Data
Diagnosing taste and odor problems : source water and treatment field guide / edited by Gary A. Burlingame ; authors: Stephen D.J. Booth ... [et al.].
    p. cm.
  Includes bibliographical references and index.
  ISBN 978-1-58321-824-2
  1. Water--Purification--Taste and odor control. I. Booth, Stephen, 1969- II. Burlingame, Gary A. III. American Water Works Association.
  TD457.D53 2011
  628.1'6--dc22

                                    2010050078

ISBN # 1-58321-824-6
      978-1-58321-824-2

American Water Works Association

6666 West Quincy Avenue
Denver, CO 80235-3098
303.794.7711

Table of Contents

Chapter 5. Making Treatment Decisions   45

Bibliography   55

Appendix A   59

Appendix B   83

Index   103

List of Figures

List of Tables

Acknowledgments

The authors acknowledge the Water Research Foundation's support of taste-and-odor control research since the early 1980s, without which this book could not have been written. The authors thank Dr. Andrew Whelton for his review.

This book is a companion book to the following AWWA resources:

Algae: Source to Treatment (2010) AWWA Manual M57, 1st ed.

Taste and Odor: An Operator's Toolbox (2001) DVD

Taste at the Tap – A Consumer's Guide to Tap Water Flavor (2010) by G.A. Burlingame

Water Quality Complaint Investigator's Field Guide (2004) W.C. Lauer

Nuisance Organisms: The Basic Facts (1997) DVD

Problem Organisms in Water: Identification and Treatment (2003) AWWA Manual M7, 3rd ed.

Introduction

The authors of this book have more than 100 years of combined experience with taste and odor (T&O) problems in drinking water. This collective experience enabled the authors to prepare this manual for operators and water system managers as a guide to better manage and respond to T&O problems.

One reason for the ineffective management and control of T&O may be a lack of understanding of the parameters to be controlled—parameters determined by the human senses. Most operators, managers, or laboratory analysts have not been trained to measure T&O.

Flavor is a term used to describe the experience when water is taken into the mouth. It consists of three separate sensations: *taste*, *odor*, and *feelings*. *Aftertaste* is the sensation that lingers in the mouth after water is swallowed. There are five taste sensations: sour, sweet, salty, bitter, and umami (a Japanese term associated with the meaty taste of proteins that is not well known in North America, nor expected to occur in drinking water).

The odor of water involves smell and feeling sensations in the nose. Odor qualities are innumerable and are typically characterized by past associations (e.g., rose, onion, rubber tire, ammonia). A taste or odor can be described by terms such as *metallic* or *musty*, and by an intensity rating such as *weak* or *strong*.

The sense of smell is the experience of odor. Odor is perceived either when substances enter your nasal cavity directly through the nostrils or when released by mouth movements during tasting and swallowing. The latter mechanism is called *retronasal* smell. Health problems like colds can block airflow and inhibit odor detection. Smell disorders can be caused by infections, disease, trauma, chemical exposure, and natural aging. Most people have some inabilities to smell certain chemicals, and some people have totally lost the ability to smell.

Feelings can be triggered by chemicals in your nose or mouth due to stimulation of your trigeminal nerve, whose endings are scattered throughout the nose, mouth, pharynx, larynx, and on the tongue. Such feelings include pain, hot and cold, pressure, and mouthfeel (i.e., astringent, drying, slick) caused by irritating substances such as ammonia, horseradish, onion, chili pepper, menthol, carbonation, and high salt concentrations.

There are several reasons why operators and water system managers may not respond to T&O problems effectively and efficiently. T&O problems often occur in unpredictable ways, such as a chemical spill. T&O problems come and go, perhaps lasting a week or less, such as during a reservoir release or after a storm. In addition, when T&O problems begin with customer complaints, they can be difficult to define and measure because customers' descriptions can be confusing or misleading. Because T&O problems are aesthetic issues and not typically public health or regulatory in nature, the motivation to resolve T&O problems can be insufficient, even though T&O problems may require changes in treatment (e.g., increased chlorination) or the addition of new treatment [e.g., powdered activated carbon (PAC)].

When a T&O problem arises, substantial data can be collected quickly to determine whether the problem is getting better or worse, and whether treatment is helping. Observations may include the appearance of a surface scum on the source water or the report of an industrial fire in the watershed. Caution and experience are required to assemble and analyze the observations to determine what is happening in order to make good decisions and avoid wrong associations or to follow "red herrings."

The authors have received many calls for help with T&O from water system operators and managers and their consultants. In many cases, the first responses failed to collect the information needed to provide effective guidance. The following is a list of a few examples:

- The collected samples were not representative of the T&O problem.
- Insufficient sample was collected to perform a complete analysis before the problem disappeared.
- No additional data were captured, such as algae type, chlorine residual, or dissolved iron concentration, to develop an association.
- The description of the problem remained vague (e.g., "water smells bad").

When a T&O problem begins, good information must be collected in a systematic way such that effective guidance can be given to identify the cause of the problem. Thus, the main purpose of this book is to provide that guidance.

Chapter 1

Sources and Causes of Tastes and Odors

Categorizing Tastes & Odors

Scientific studies have revealed many causes for T&O problems; thus the descriptions of tastes and odors have been categorized. The categories evolved from a carefully detailed evaluation of the technical literature pertaining to T&O problems and treatment chemistry.

Some water treatment facilities still use outdated classification schemes to describe odor (Table 1-1).

In general, four broad categories exist for "off" flavors in drinking water. These categories are defined by the sources of odorants and general environmental conditions. The categories in Table 1-2 do not include tastes that could occur in water (salty, sour, bitter) or mouth-feel sensations (oily, drying, astringent, chalky).

A T&O wheel that helps provide detailed categorization is published in *Standard Methods for the Examination of Water and Wastewater* under *2170 Flavor Profile Analysis*.

Flavor profile analysis (FPA) was established as a standardized sensory testing method for understanding taste and odor. The flavor of a water sample may be described as "moderately musty, slightly grassy with a barely noticeable chlorinous odor."

Water temperature and pH are important factors that affect water's perceived flavor. Tap water's serving temperature can range from 4°C to about 40°C. Cold water has a flavor that can be difficult to describe, and hot water can burn the mouth. Water's pH also impacts taste and is best between 6.5 and 8.5. Below pH 6.5, water can taste bitter, and above pH 8.5, water can feel slippery.

Much research has been performed in the science of aesthetic water quality, but utilities have had difficulty incorporating this science into their practices.

Table 1-1   Example of Odor Description Categories from 1948 that provided a basis for many subsequent schemes

a = aromatic	G = geranium	o = oily
c = cucumber	I = iodoform, medicinal	p = pigpen
Cl = chlorinous	Mm = muskmelon	s = sweetish
e = earthy	m = moldy	S = hydrogen sulfide
f = fishy	M = musty	v = vegetable
g = grassy	N = nasturtium	V = violet

Adapted from The Microscopy of Drinking Water. 1948.

Table 1-2   Broad Categories of Odors in Water

I.	Naturally Occurring Odors
	Grassy/hay/straw/woody
	Fragrant/vegetable/fruity/flowery
II.	Industrial or Manmade Odors
	Chemical/hydrocarbon/miscellaneous
	Medicinal/phenolic
	Chlorinous/ozonous
III.	Microbiologically Produced Odors under Aerobic Conditions
	Earthy/musty
	Earthy/potato bin
	Tobacco-like
	Fishy
IV.	Microbiologically Produced Odors under Anaerobic Conditions
	Marshy/swampy/septic/sulfurous
	Decaying vegetation
	Fishy/rancid

Chlorine—The Most Common Odor in Drinking Water

In North America, it is a common practice, and typically a regulatory requirement, to provide a measurable level of chlorine residual (as free chlorine or chloramine) at the taps of water consumers. Thus drinking water typically has a chlorinous smell or flavor. The average consumer is more sensitive to free chlorine [starting at 0.5–1.0 milligrams per liter (mg/L)] than monochloramine (starting at 3 mg/L), even though a trained sensory panel can detect these chemicals at

Table 1-3   Flavor Profile Analysis of Surface Water Source and Finished Drinking Water

T&O Characteristics	Treatment Plant on Delaware River		Treatment Plant on Schuylkill River	
	Times Detected (Percent)	Average Intensity by FPA*	Times Detected (Percent)	Average Intensity by FPA*
Source Water Odors				
Decaying Vegetation	85	4	81	3
Septic	45	3	23	2
Earthy	30	3	40	3
Fishy	21	3	4	2
Drinking Water Odors				
Chlorinous	86	2	98	4
Musty	74	2	63	2
Earthy	9	2	3	2
Drinking Water Flavors				
Chlorinous	67	2	95	2
Musty	55	2	52	2
Earthy	13	2	7	2

* FPA flavor intensity scale of 1–12 (just detectable, very weak to strong)

much lower levels in water. The descriptions used to describe the flavor of chlorine residuals are "chlorinous," "swimming pool," and "bleachy." The presence of a chlorine odor in drinking water can sometimes mask or interfere with the consumer's detection of other odors, such as that of geosmin, an earthy smelling chemical.

When chloramine is used, the goal is to ensure that the predominant species is monochloramine. In the 6–7 pH range, 20 percent of the residual total chlorine can be dichloramine. In the 5–6 pH range, 40 percent of the residual total chlorine can be dichloramine. Dichloramine has a stronger chlorinous odor than monochloramine.

As a general rule of thumb, dichloramine should represent less than 20 percent of the total chlorine residual to prevent objectionable chlorinous odors. A chlorine-to-ammonia ratio between 3 and 5 to 1, and a pH between 7.5 and 8.5, are recommended. Some Midwestern lime-softening treatment plants that use chloramine as the secondary disinfectant maintain a finished water pH of 9 or above and produce an essentially odor-free water.

Tastes Caused by Different Source Waters

Consumers expect their drinking water to be pleasant and refreshing, and the combination of tastes and odors in the water together produce a "flavor" that is acceptable to the consumer. Some people, however, cannot taste anything; their condition is called *aguesia*. This section will focus on the tastes of water—sweet, salty, sour, and bitter.

Individual preferences for mineral content in water vary. Some people like waters with high mineral content; others do not. In general, people like to drink the water they are used to drinking. This is similar to "brand loyalty"; as with other consumer products, people become familiar with certain waters.

Minerals enter natural waters by weathering, erosion, or disturbance of rock and soil, but they can also come from anthropogenic sources such as road salt and industrial discharges. The mineral content of natural water may be altered by water softening or reverse osmosis such that if deionized water (such as a laboratory MilliQ water or water desalted by reverse osmosis) flows over a tongue, there may be an unpleasant, persisting bitter aftertaste.

Total dissolved solids (TDS) is a measure of mineral content, which includes common cations such as calcium, magnesium, potassium, and sodium as well as carbonate, bicarbonate, chloride, nitrate, and sulfate anions. The United States and Canada limit TDS to a maximum of 500 mg/L, while the World Health Organization established 1,000 mg/L TDS as its guideline. Acceptable TDS concentrations vary according to population preferences: high TDS would be 251 to 500 mg/L; moderate, 101 to 250 mg/L; and low, less than 100 mg/L. Maintaining TDS at less than 250 mg/L is advisable.

Taste buds are not equally sensitive to all ions. For instance, a person may be very sensitive to sodium (Na^+) but less sensitive to the tastes of calcium (Ca^{2+}) and potassium (K^+). Thus, it takes a higher concentration of calcium to elicit the same amount of taste as a small amount of sodium. Good tasting water has been reported to have a total hardness between 10 and 100 mg/L as $CaCO_3$, mainly because of the presence of calcium. Calcium tends to be the major cation present in most waters. Magnesium typically occurs at lower concentrations in natural waters than calcium, but it is more readily tasted compared to calcium. Too much calcium or magnesium can cause a bitter taste as well as a mouthfeel. Soft waters with their low levels of calcium and magnesium tend to have no effect on taste, but the absence of minerals in water can produce a "slick" feeling on the tongue. Customers may complain about the hardness of water, depending on the water quality they are accustomed to. A water supply that alternates between soft and hard water could initiate customer complaints simply because of the change in hardness.

Untreated groundwaters can taste salty. Waters with a high mineral content tend to impart a salty taste, especially when sodium chloride is present at or above 500 mg/L, a concentration that can be reached after salt water intrusion. Sodium, which produces a salty taste, is commonly found in water, although most drinking waters contain less than 50 mg/L. The USEPA recommends that sodium not exceed 30 to 60 mg/L to prevent any taste effects and to help in reducing the sodium intake for users on a low-sodium diet.

Trace metal cations, such as iron and copper, are very flavorful in water. The threshold concentration for ferrous iron is around 0.05 mg/L; however, some people can detect the metallic flavor of ferrous iron at concentrations as low as 0.005 mg/L. Ferrous iron is common in groundwater, which tends to lack oxygen that would oxidize the ferrous iron to ferric iron.

Copper is typically present at microgram per liter (ug/L) concentrations in natural waters, but it can also leach from pipes and other plumbing fixtures into drinking water. People may detect the metallic or bitter taste of soluble copper at 0.5 mg/L, although the threshold for detecting copper ranges from less than 0.1 to more than 10 mg/L.

Anions such as bicarbonate, chloride, and sulfate can impact taste. Together, bicarbonate and carbonate are the major ions contributing to the alkalinity or buffer capacity of water. The taste of bicarbonates, which occurs at pH values less than 8.3, is preferred over carbonates. Most tap waters contain less than 150 mg/L of bicarbonate.

Chloride can be tasted at concentrations above 200 to 300 mg/L. Increased chloride levels in water, in the presence of sodium, calcium, potassium, and magnesium, can cause an objectionable taste. Sulfate has minimal taste impact in low mg/L levels as typically found in most waters. Above 250 mg/L, sulfate can impart a salty taste.

Odors Caused by Source Waters

A common natural cause of odors in groundwater is hydrogen sulfide from sulfate-reducing bacteria, causing an obnoxious rotten-egg odor. However, hydrogen sulfide can also be produced in the anoxic zones of a reservoir, where sulfate-reducing bacteria reduce sulfate to hydrogen sulfide. Anthropogenic sources of groundwater contamination have resulted in the common contaminants bromodichloromethane, chloroform, methyl-*tert*-butyl-ether (MTBE), perchloroethene (PCE) and trichloroethene (TCE). MTBE has been of particular concern for its impact on the odor quality of water.

While groundwater is typically very low in odor or has no odor at all, surface water tends to contain background odors, including such descriptions as earthy, musty, grassy, green vegetation, marshy, decaying vegetation, and fishy. The decay of algal biomass or terrestrial plants can result in the development of decaying vegetation, rotten egg and septic odors caused by alkylsulfides. Sulfur bacteria can break down proteins and amino acids (the building blocks of life) to organic sulfides, such as dimethyl-sulfide, dimethyl-disulfide and dimethyl-trisulfide, that give decaying vegetation and septic odors.

The most common odors caused by the proliferation of algae are earthy and musty, associated with the release of geosmin and 2-methylisoborneol (MIB), respectively. Geosmin and MIB are very similar and are found worldwide. They are two of the most common odor problems for drinking water utilities. Geosmin has an earthy, dry dirt, corn silk, beet type of odor that occurs at levels between about 2 to 200 nanograms per liter (ng/L). MIB has an earthy, wet mulch, peaty type of odor that occurs at levels between about 2 to 100 ng/L. Sometimes they co-occur. Both are produced by cyanobacteria. Analytical detection methods can measure below 5 ng/L. Their odor threshold concentrations are around 5-10 ng/L. Water utilities have found that by keeping the drinking water levels of geosmin and MIB from exceeding about 10 ng/L, customer complaints are kept to a minimum.

Cyanobacteria, chlorophytes (green algae), and diatoms (Fig. 1-1) have been shown to produce various odors, such as grassy, fishy, medicinal, and cucumber. The latter odor is caused by the release of an aldehyde, *trans*-2, *cis*-6-nonadienal.

Man-made chemicals are an important cause of odors in surface waters. The most common descriptors belong to the chemical, hydrocarbon, miscellaneous category. The chemicals involved include hydrocarbons, gasoline additives (MTBE, ethyl *tert*-butyl ether), dioxanes and dioxolanes (from polyester resin manufacturing), and solvents.

Odors Caused by Treatment Processes

The most common unplanned odors generated during water treatment fall into the fruity/fragrant and medicinal odor categories. Orange-fruity odors can be caused by the aldehydes formed during

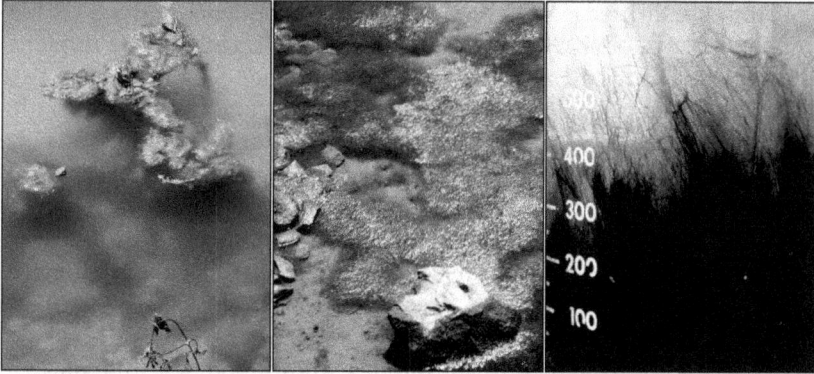

Courtesy of Gary Burlingame.

Figure 1-1 Examples of algae growth as would appear in surface waters

ozonation. These odors are easily removed by subsequent granular activated filtration and biological degradation.

However, when chlorine is applied to water already laden with algae, it can cause the subsequent release of odorous metabolites (such as geosmin and MIB) at high concentrations. Chlorination of algae can also produce aldehydes such as hexanal, heptanal, and nonanal, which impart an "off" flavor when their combined concentrations exceed 1 to 10 ug/L.

Free chlorine and residual chlorite can react to reform chlorine dioxide in the distribution system, which imparts a chlorinous odor, according to the following reaction:

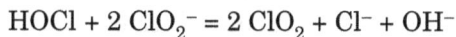

$$HOCl + 2\ ClO_2^- = 2\ ClO_2 + Cl^- + OH^-$$

In addition, volatilized chlorine dioxide can react with volatile organic chemicals (VOCs) from new carpet (or other sources) inside customers' homes to create kerosene-like and cat urine-like odors.

Other odors generated during treatment can result from the reaction of chlorine disinfectants (chlorine, monochloramine) with man-made or natural contaminants such as phenols. When phenols are present in the source water, their reaction with chlorine leads to the formation of chlorinated phenols and, in the presence of bromides, leads to brominated and mixed chloro-bromo phenols.

While the phenolic precursors may remain unnoticed because of their high odor threshold (above 1 mg/L for phenol), some of the by-products formed, in particular 2,6-dibromophenol and 2,6-dichlo-rophenol, have odor thresholds in the low ng/L range. The unpleasant odor is described as medicinal/chlorophenol. The formation rate of halogenated phenols is lower with monochloramine than with chlorine. Chlorine dioxide does not yield either chloro- or bromo-phenols.

Ozone is often associated with better tasting water. However, ozone use can form fruity odors from aliphatic aldehydes such as heptanal, nonanal, and ketones with carbon chains of 3 to 10 carbon atoms (C_3-C_{10}). Ozone also causes oxidant odors and can produce medicinal odors in the presence of bromide and iodide. (See Figure 1-2.)

When waters containing bromides and/or iodides and natural organic matter are chlorinated, brominated and iodinated disinfection by-products can form, which result in medicinal or pharmaceutical "off" flavors. When such waters are ozonated, similar objectionable odors are caused by the formation of bromo/iodo-cresols or phenols.

Iodides are naturally present in many groundwaters in coastal areas or ancient sedimentary deposits. The reaction of iodides with chlorine and especially with chloramine results in the formation of iodo-trihalomethanes (iodo-THMs), which impart a medicinal odor at low µg/L levels (e.g., 1 µg/L for iodoform). Chlorine dioxide does not yield iodo-THMs. An alternative remedial action is to use a higher chlorine dose, which leads to the formation of nonodorous iodates rather than iodo-THMs. In the case of chloramination, adding chlorine to the water 30 minutes or so before ammonia is added to the water minimizes the formation of iodo-THMs and medicinal odors. Both phenolic and iodide precursors can be destroyed by ozonation prior to the application of the chlorine.

Another source of odors during treatment can be the sludge generated in lime softening as well as in conventional treatment processes. If allowed to accumulate and remain at the bottom of a sedimentation basin, microorganisms can proliferate in the anoxic or anaerobic sludge. Reduced sulfur compounds may form within the sludge and cause rotten egg, sulfide, or septic odors. This condition can also result in the mobilization of metals back into the water. The aesthetic quality of water reclaimed from sedimentation basin sludge or backwash wastes should be evaluated to prevent odors from those potential sources.

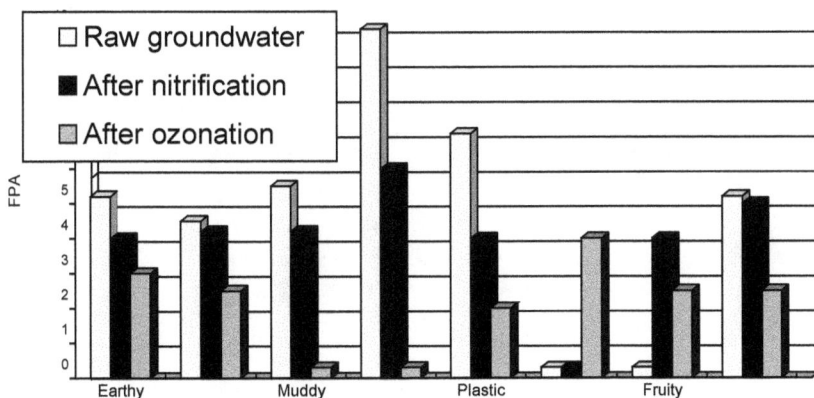

Courtesy of Journal AWWA.

Figure 1-2 Example of how odor quality (by Flavor Profile Analysis odor intensity) can change through a treatment process

Chapter 2

Requesting Lab Tests During a T&O Problem

Tips on Sampling

An understanding of good sampling techniques ensures that the analytical data produced in the laboratory can be related back to the condition of the water that was sampled. First, make sure that the sampling supplies are appropriate. Samples should be collected in clean 500 mL or 1-L glass bottles (i.e., borosilicate) that are equipped with either Teflon-lined caps or caps made from clear PVC. Improper bottles and caps can impart trace levels of odorants to the water. The bottles can be purchased directly, or the laboratory that will perform the analyses can supply them. Bottles should come in dedicated coolers with bubble wrap or similar protection, and once collected should be placed out of direct sunlight. Amber bottles are recommended to minimize the effect of sunlight. Samples that cannot be refrigerated within 2 hours after collection should be stored in insulated ice chests containing ice packs made from nonresilient, hard-plastic materials. Soft ice packs with gels can develop fishy odors.

Always collect more than is needed when diagnosing a T&O problem. Samples can be stored and disposed of later if additional analyses are unnecessary. However, once a T&O problem subsides, samples can no longer be collected to perform additional analyses, and the problem will remain undiagnosed until it returns.

The sample should be collected from the treatment process or pipe, not from the plumbing of the building or some other unrepresentative sampling line. The line should be well flushed before sampling to prevent the collection of a sample that represents local plumbing problems rather than conditions in the water supply.

When sampling from a tap, all screens and aerators should be removed to minimize turbulence. The sample tap's flow rate should be reduced during sampling. The sample bottles should be rinsed three times with the sample water, then slowly filled until they are headspace-free and capped. The laboratory should be provided with a thorough description of sample location, date, time, and sampler contact information in case the laboratory has questions later.

A scuba diver may be needed when sampling from reservoirs. Cyanobacteria can grow in specific locations on the bottom of a reservoir, on walls and submerged matter, or at different depths. The only way to obtain a sample may be to dive for it.

If samples are collected from bridges or elevated areas above waterways or if samples are collected at different depths in a water body, special sample collection devices should be used. Vertical and horizontal samplers (*Kemmerer* bottles and *van Dorn* bottles, respectively) can be lowered into a water body to collect samples at various depths. These must be thoroughly precleaned and rinsed between samples to reduce carryover. Filling the bottles with odor-free (deionized) water between samples and collecting these rinsates for analysis is one way to provide some quality control and to ensure that carryover from one sample to another does not occur.

Summary of T&O Test Methods

Water utilities should incorporate one or several T&O evaluation methods into their routine water-quality monitoring program. The methods complement chemical and biological monitoring. The most attractive features of sensory methods are their simplicity and potential applications to a wide variety of situations, such as detecting subtle changes in source and treated waters, evaluating the effectiveness of treatment processes, and tracking odors to their sources. Adoption of these methods may require funding for staff education, odor ability screening, and training through hands-on workshops given at water industry conferences or at individual utilities.

Just as no single analytical chemistry method can detect all chemical contaminants in water under every given set of conditions, no single sensory method can provide all answers to T&O questions. The water industry has a robust toolbox of methods to evaluate sensory properties of water, including simple, yet reliable methods for routine daily monitoring by water treatment plant personnel.

Sensory methods have been used to evaluate foods and beverages since the early 1900s and can be applied successfully to drinking water in the following ways.

- Routine monitoring of source or finished water to detect aesthetic changes over time
- Evaluating treatments designed to remove tastes or odors

- Tracking T&O problems in watersheds and distribution systems
- Evaluating water samples from customers' premises
- Early warning of the occurrence of T&O in source waters
- Assessing odors from distribution materials in contact with water

To control tastes and odors, it is crucial to locate their source: the water resource, the treatment plant, the distribution network, or customer plumbing. Simple tests can be performed to identify the source of a problem.

The following types of containers are recommended for testing, in order of preference (Figure 2-1):

- One-liter glass flasks with glass stoppers
- One-liter amber flasks
- Commercially available water glasses

Testing should take place in a low odor area such as an office or meeting room and should be carried out with actual glass cups rather than plastic cups. The glass must be rinsed with bottled water between samples. It is also important for the tasters not to have drunk coffee or smoked for at least 30 minutes before the testing tests. The tasters must not have used perfume or deodorant and must not suffer from a cold or allergy.

For disinfected waters, the sample in the glass must be sniffed first, then tasted in the mouth. For nondisinfected samples, it is important to assess the sample by sniffing only, to qualify common odors in particular.

A report sheet must be available to record the results for each sample (presence or absence of T&O, tentative description.)

One or two people can monitor water for unusual T&O, but five people may be needed to confirm when a taste or an odor has been detected.

To evaluate whether an analyst has a normal sense of smell, commercial products can be purchased from the following sources:

- Smell Identification Test: Sensonics, Inc., 125 White Horse Pike, Haddon Heights, NJ 08035; phone: (800) 547-8838 or (856) 547-7702; www.smelltest.com
- Sense of Smell Kit: Carolina Biological Supply Company; 2700 York Road, Burlington, NC 27215; phone: (336) 584-0381; www.carolina.com

These products are designed to screen the senses for a normal sense of smell, not to screen for odorants specific to drinking water. These odor kits are valid for:

- Understanding how the senses work
- Testing general smell ability

Sensory methods used in the water industry are presented in the following sections and organized into categories according to the purpose of the test. It is important to remember *safety first*, and panelists should not assess samples that could contain harmful microbiological or chemical agents.

Flavor Rating Assessment (*Standard Method 2160*). Flavor Rating Assessment is an acceptance test that uses a nine-point scale with four to eight trained panelists or consumers to estimate the acceptability of water for daily consumption. Its purpose is to assess a person's "liking" for a drinking water. Each analyst determines the acceptable level of odor-causing or taste-causing substances in water at room temperature, and the water is rated by averaging the ratings of each analyst. The scale ranges from 1 to 9:

1. I would be very happy to accept this water as my everyday drinking water.

2. I would be happy to accept this water as my everyday drinking water.

3. I am sure that I could accept this water as my everyday drinking water.

4. I could accept this water as my everyday drinking water.

5. Maybe I could accept this water as my everyday drinking water.

6. I don't think I could accept this water as my everyday drinking water.

7. I could not accept this water as my everyday drinking water.

8. I could never drink this water.

9. I can't stand this water in my mouth and I could never drink it.

Flavor Profile Analysis (*Standard Method 2170*). Flavor Profile Analysis (FPA) is the "gold standard" for drinking water T&O analysis. It uses a trained panel of four to seven analysts to describe T&O attributes and assign intensity ratings. Any taste or odor—from earthy-musty to chlorinous to sweet floral—can be assessed by this method. The method requires intensive training and standardization in order to achieve reliable and reproducible results.

Courtesy of Andrea Dietrich and Gary Burlingame.

Figure 2-1 Various types of odor testing supplies

Samples are evaluated either at room temperature (25°C) or after being warmed to 45°C. Flavor is evaluated at room temperature.

Each assessment provides one or more descriptors and an intensity rating for each on a scale of 0 to 12, with 12 being the strongest. After evaluating the sample, analysts discuss their individual opinions. If at least 50 percent agree on a descriptor, an average intensity is calculated. An FPA panel can identify more than one taste or odor component in a water sample, thus generating a "profile" of the sample's flavor character.

The first odor standard developed for the drinking water industry was n-Hexanal. It has a grassy, green apple, pumpkin odor and is frequently found in water supplies as a result of vegetation. A concentration of 4 µg/L (in water) yields an FPA intensity of 4 at 45°C.

Attribute Rating Test (*Practical Taste-and-Odor Methods for Routine Operations: Decision Tree,* Dietrich et al., 2004, Water Research Foundation). The Attribute Rating Test (ART) is a paired-comparison test by which one identifies specific odorants (e.g., geosmin or MIB) in water samples warmed to 45°C. This test rates the odor intensity as less than, equal to, or greater than that of standards containing a known concentration of the odorant. For geosmin or MIB, 15 ng/L is used because it is the concentration at which many consumers begin to register complaints. ART is an efficient, focused, cost-effective method that treatment plant personnel can use to monitor raw and treated water for specific odorants and to maintain the finished water's odor below a target concentration.

2-of-5 Odor Test (*Practical Taste-and-Odor Methods for Routine Operations: Decision Tree,* Dietrich et al., 2004, Water Research Foundation). This test determines whether two samples are similar or different. Five flasks are warmed to 45°C and randomly mixed, then placed on a rotating plate such as a lazy Susan and twirled. The analyst then sniffs the samples and sorts them, based on their odor characteristics, into one group of two and another group of three. If the samples are sorted correctly, the two waters are considered to be noticeably different. Finally, the analyst describes the difference by selecting a term from a list of common odor attributes.

The test is an excellent quality-control test, as it allows panelists to detect very small differences between the odors of samples and the control. Differences in the odors signal the development of problems. The power of this test method is its simplicity. For example, customer complaints could be compared to the water leaving the treatment plant to see if the T&O problem is present in both waters.

Threshold Odor Number (*Standard Method 2150*). The Threshold Odor Number (TON) test is designed to assess the number of dilutions with odor-free water necessary to eliminate the odor in a water sample. The procedure uses odor-free water to dilute the water sample to be tested; the odor is evaluated at either 60°C or 40°C. The total volume (odor-free water volume + sample volume) to be used in the test is 200 mL. The TON is the number obtained by dividing the total volume by the volume of the odorous sample being evaluated. Thus, if the first dilution in which the odor could be detected contained 5 mL of the sample, the TON would be (195 mL + 5 mL) divided by 5 mL, or TON = 40. The TON is not a measure of odor intensity, but a measure of an odor compound's persistence upon dilution.

Flavor Threshold Test (*Standard Method 2160*). Flavor Threshold Test (FTT) is the application of the ASTM method of limits to drinking water flavor or odor. It estimates the sensory threshold of detection for odorants in water using only one ascending series of presentations that increase in concentration of the odorant. FTT recommends screening assessors based on sensitivity to n-butyl alcohol and o-chlorophenol, both of which have well established thresholds. An average threshold is calculated at the first point of detection from many assessors.

American Society for Testing and Materials, Threshold Testing (*ASTM E679-91*). ASTM Method E679-91, Standard Practice for Determination of Odor and Taste Thresholds by a Forced-Choice Ascending Concentration Series Method of Limits, is a test for estimating sensory thresholds that addresses both detection and recognition thresholds. The method uses a series of samples from low to high concentrations selected to bracket the expected threshold level. Each panelist performs a triangle test (one sample with the odor in it and two without) at each concentration. From the data of correct responses, an individual's odor threshold is estimated, and a group estimate is calculated using the geometric mean of individual thresholds.

In selected situations, T&O evaluation methods can be considered as a screening test to determine whether water may have a problem. The 2-of-5 test, triangle test, and rating method for evaluating distribution-system odors in comparison to a control are well designed for use as screening tests. For example, if a customer complains of an "off" odor in tap water, a utility could compare the customer's sample to the plant effluent using any of the abovementioned tests as a first step to determine whether a difference exists. If no difference is found, then the odor of the customer's water is not different from that of the water leaving the treatment plant. If a problem is found, the utility can perform additional chemical or biological testing to find the source of the difference.

Testing for Chlorine or Chloramine

All treatment plants or water supplies that disinfect water with chlorine or chloramine already sample and test their water for chlorine residual. This can be reported as total chlorine residual, free chlorine residual or combined chlorine residual.

Chloramine chemistry is complicated. Chloramine can exist as monochloramine, dichloramine, or trichloramine. Dichloramine is typically present to some degree when a monochloramine residual is used. The dichloramine residual, though, should be kept as low as possible (less than 0.2 mg/L as chlorine) to minimize its impact on the chlorinous flavor of the drinking water. Samples should be sent to a laboratory to analyze for dichloramine. Trichloramine is not usually found and should be kept below detection, as it has the strongest odor of the three chloramine species. The combined chlorine residual can be measured using approved field test kits.

Water utilities that use chloramine should also test for ammonia residual. A free ammonia residual that exceeds 0.5 milligrams nitrogen per liter (mg-N/L) may affect the flavor or odor of the water. If the source water contains high levels of ammonia, even though the water supplier may not intentionally feed a chloramine residual, testing should be conducted to understand the chlorine chemistry in treatment and in the finished drinking water. Ammonia field test kits can be used, but the results should be checked periodically against laboratory analysis, especially when unusual results are obtained.

Testing for Inorganic Chemicals

Inorganic contaminants, such as chloride, manganese, and iron, can travel through the treatment process. They can originate from untreated groundwater (e.g., iron, manganese, and sulfate) or directly from chemical treatment [e.g., manganese (potassium permanganate), iron (iron-based coagulants), and aluminum (alum)].

The USEPA's Secondary Maximum Contaminant Levels (SMCLs) (Table 2-4) are a good place to start when considering which inorganic analyses to perform and to understand at what levels these contaminants need to be tested. As stated in the Preface of the USEPA 1979 publication *National Secondary Drinking Water Regulations*:

> *Secondary Maximum Contaminant Levels (SMCLs) are established for chloride, color, copper, corrosivity, foaming agents, iron, manganese, odor, pH, sulfates, total dissolved solids and zinc. At considerably higher concentrations, these contaminants may also be associated with adverse health implications. These secondary levels represent reasonable goals for drinking water quality, but are not federally enforceable. Rather, they are intended as guidelines for the State....States are encouraged to implement these SMCLs so that the public will not be driven to obtain drinking water from potentially lower quality, higher risk sources.*

The US Environmental Protection Agency (USEPA) *Secondary Drinking Water Regulations: Guidance for Nuisance Chemicals* (July 1992) lists the following three categories of effects that can be associated with the SMCLs:

1. Aesthetic (odor, taste, color, foaming)

2. Cosmetic (skin discoloration, tooth discoloration)

3. Technical (corrosion, scaling, sedimentation)

Laboratories should analyze for SMCL parameters at the levels of concern. Laboratories will provide sample bottles with or without preservatives, as well as sampling instructions, if needed. Most of these tests are relatively simple and low in cost. Field test kits are available for pH, chloride, copper, iron, manganese, and sulfate at

Table 2-4   USEPA Secondary Maximum Contaminant Levels (SMCL) for Chemicals that Can Affect the Taste of Drinking Water (EPA 1979)

Contaminant	SMCL (mg/L)	Effects
Aluminum	≤ 0.05-0.2	Colored water
Chloride	≤ 250	Salty taste
Color	≤ 15 color units	Visible tint
Copper	≤ 1.0	Metallic taste, blue-green staining
Corrosivity	Noncorrosive	Metallic taste, corrosion, fixture staining
Foaming agents	≤ 0.5	Frothy, cloudy, bitter taste, odor
Iron	≤ 0.3	Rusty color, sediment, metallic taste, reddish or orange staining
Manganese	≤ 0.05	Black to brown color, black staining, bitter metallic taste
pH	≥ 6.5 ≤ 8.5	Low pH – bitter metallic taste, corrosion High pH – slippery feel, soda taste, deposits
Sulfate	≤ 250	Salty taste
Total Dissolved Solids	≤ 500	Hardness, deposits, colored water, staining, salty taste
Zinc	≤ 5	Metallic taste

the sensitivity necessary to screen waters for possible problems. Laboratory analyses should be used on occasion to confirm the results obtained using field test kits.

Sulfides can be a concern, especially for groundwater, and can cause customer complaints in the distribution system. Sulfides can exist as hydrogen sulfide or as other inorganic or organic sulfides. Field test kits are available for sulfides, but their results should always be confirmed with laboratory analyses. Careful sampling and handling instructions may be needed, depending on the type of sulfide to be analyzed. Rotten egg odors from hydrogen sulfide (odor threshold estimated at 50 to 100 µg/L in water) are analyzed by gas chromatography. Test kits are also available for hydrogen sulfide.

Laboratory analyses do not require large volumes of samples. In addition, analyses for metals (e.g., aluminum, copper, iron, manganese, zinc) can all be performed from a single sample.

Testing for Organic Chemicals

The challenge with conducting organic chemical analyses is two-fold. First, chemicals present at very low levels in water can cause T&O. Second, the background matrix of inorganic and organic matter can interfere with the detection of the chemicals of interest. Therefore, samples must be collected and handled carefully. Compounds of interest must be extracted and separated from the background matrix and then concentrated so that the laboratory instrument can quantify them with some level of certainty.

The investigation of odor from organic chemicals often requires the use of gas chromatography (GC), usually coupled with mass spectrometry (MS) for detection and identification. Some of the most common odor descriptors, their usual causative agents, the target limits of detection (LOD) compatible with odor thresholds and suitable analytical methods are summarized in the Table 2-5.

Solid phase micro-extraction (SPME) has become the method of choice for earthy-musty odors caused by geosmin and MIB. An LOD of 1 to 2 ng/L can be achieved, which is below the odor threshold of these two compounds (about 5 ng/L). Lower limits of detection are necessary if trichloro- or tribromo-anisoles are involved. The stir-bar-sorptive extraction (SBSE) method, a popular technique in

Table 2-5   Organic Chemicals that Can Cause Odors

Odor Descriptor	Target compounds	Target concentrations
Earthy-Musty	Geosmin	1 ng/L
	MIB	
	2,4,6-trichloroanisole	50 pg/L
	2,4,6-tribromoanisole	50 pg/L
Medicinal	Iodoforms	0.5 µg/L
Chlorophenolic	Chlorophenols	5 ng/L
	Bromophenols	5 ng/L
Hydrocarbons	Hydrocarbons	1-10 µg/L
Solventy	Solvents	1 µg/L
Septic-Swampy	Dimethyltrisulfide	10 ng/L
	Dimethyldisulfide	200 ng/L
Fishy	Heptadienal	2.5 µg/L
Cucumber	t-2, c-6-Nonadienal	10 ng/L
Fruity	Aldehydes>C6	0.1 µg/L
Olive oil–Tutti frutti	Dioxanes and dioxolanes	10 ng/L

Europe but less common in North America, can be combined with GC-MS to achieve an LOD of 0.1 ng/L. Another possibility is to combine closed-loop-stripping-analysis (CLSA) with a large volume injection of the sample extract in GC-MS. This allows LODs around 20 picograms per liter (pg/L) to be achieved.

Chemical odors (solvent, shoe-polish, varnish, rubber, glue, plastic, turpentine) may come from a variety of causative agents. It is recommended to combine analytical methods for both volatile and semi-volatile compounds, such as P&T (purge and trap) with GC-MS together with LLE with GC-MS and CLSA; or SPME or SBSE with GC-MS.

Medicinal odors caused by iodo-THMs can be diagnosed using the traditional CLSA-GC-MS approach. Although recoveries by this method are low, the concentration factor involved is high enough to detect iodoform and other mixed chloro-, bromo-, iodo-THMs (such as chlorobromo-iodomethane) whenever the typical medicinal odor is detected.

For a more quantitative analysis, LLE by methylene chloride is a preferred approach. A current trend is to use SPME-GC-MS. Chlorophenol odors caused by chlorinated or brominated phenols (2,6-dibromo- and 2,6-dichloro-phenols) have flavor thresholds around 5 ng/L. These can be detected by SPME-GC-MS or SBSE-GC-MS.

P&T or head-space (HS)-GC-MS is applicable to the analysis of volatile and semivolatile hydrocarbons (white-spirit gasolines and their oxygenated additives as well as the volatile fraction of fuel-oil or gas-oil). LLE is a better method for heavier petroleum fractions such as new or used motor oils.

Volatile sulfides have been measured in water samples. The low odor threshold of dimethyltrisulfide (10 ng/L) in water can be reached by CLSA-GC-MS. P&T and HS-GC-MS can be used for the more volatile dimethylsulfide and dimethyldisulfide.

Alkyl dioxanes and dioxolanes impart odors at low ng/L levels. North American people tend to describe the odors as "tutti-frutti," while Latin American people describe them as "olive oil-like." These odorants can be determined by CLSA-GC-MS. The same method is well adapted to the determination of trans-2, cis-6-nonadienal, an algal metabolite that yields a cucumber odor. Simultaneous distillation-extraction (SDE)-GC-MS is more efficient at extracting the more polar 2,4-heptadienal, a compound causing a fishy odor.

Testing for Bacteria

T&O problems can sometimes be related to a certain type of bacteria such as iron, manganese, and sulfur bacteria. A problem can be indicated by the biofouling of pumps, corrosion of iron materials, black or red colored water, biological or inorganic particulates, or the staining of fixtures. Direct analysis for iron and sulfur bacteria should be done by microbiologists who are familiar with these bacteria. Slime growths and bacterial masses are sometimes, but not always, indicators of such bacteria.

Monitoring for total iron, total manganese, or total sulfides, or other parameters such as pH and turbidity, may be a more feasible approach. Differentiating between total and dissolved forms of iron and manganese can indicate certain conditions. Field test kits are available for iron and manganese. The BART^{TM} test system is designed for detecting these nuisance bacteria, but grab sampling may not properly indicate the presence of biofilm in the well. Because biofilm will slough off the well on occasion, there may be no easy way to ensure a representative sample was taken for testing. Therefore test results have to be interpreted with caution.

Earthy–musty odors may occur when no algal growth is observed in the water column. Various species of actinomycetes (*Streptomyces, Nocardia, Microbispora*) may produce geosmin and MIB in laboratory cultures. Because these organisms can be washed from soils and are present in bottom sediments or inside algal cells, their presence during odor episodes is more difficult to demonstrate. These bacteria can also produce a "potato-bin" odor caused by 2-isopropyl-3-methoxypyrazine.

Actinomycetes are very difficult to measure and monitor (Figure 2-2). The culture method is difficult. Their occurrence in water could be caused by spores or vegetative fragments. Thus, detection does not indicate whether they are actively growing or merely dormant.

Testing for Algae and Cyanobacteria

T&O problems are frequently related to algae or cyanobacteria (blue-green algae). More than one alga can cause the problem. Correct data interpretation (and successful diagnosis of T&O) is critically dependent on the methods used to collect, store, and ship samples prior to analysis. Indiscriminately or incorrectly applied procedures can render costly field and lab work useless. Prolonged or incorrect storage (and freezing!) can produce significant changes in samples.

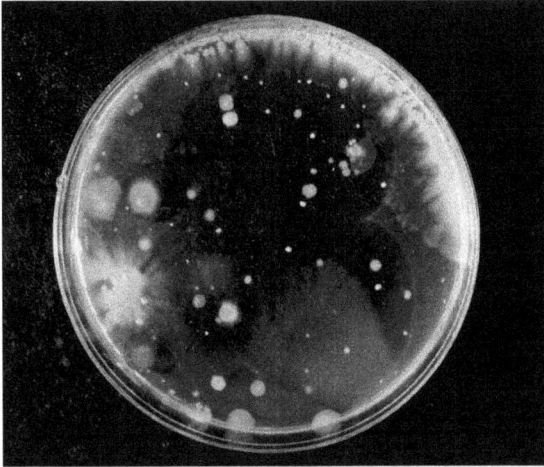

Courtesy of Gary Burlingame.

Figure 2-2 Actinomycetes growing on an agar plate

Samples should be processed and preserved rapidly because many delicate algae rapidly lyse after collection. Lugol's iodine is the most recommended means for preserving algae, particularly for planktonic samples. It is widely used and noncarcinogenic.

Field samples can be fixed onsite by pouring a well mixed subsample into a 100 mL Boston round flint glass bottle (or other suitable glass container with a polycone-lined cap) prefilled with a small volume of Lugol's to result in a final concentration of 1 to 2 percent or to a light tea color (e.g., 1 to 2 mL of Lugol's in a 100 mL sample). Lugol's-preserved samples can be stored up to a year or more in the dark at room temperature, but it is recommended that an additional seal of Parafilm be wrapped around the closure and the color checked periodically. A few additional drops of Lugol's can be added if samples begin to lose color.

It is often helpful to retain some live material because preservation destroys motility and alters color and shape. This material can be examined using a simple wet slide mount, if the resources and need exist to isolate and culture target species.

Several techniques may be used to collect, count, and identify algal field samples. Such techniques vary in the effort and expertise involved as well as the level of accuracy and taxonomic detail they deliver. Most methods are fairly time-consuming and require a level of expertise that is not available in most drinking water utilities. Therefore, most water utilities hire outside experts for algal studies.

With the development of advanced techniques in genetics, proteomics, and analytical chemistry, early species classification systems based on morphology and pigments have been revised. The revised species names can lead to confusion when comparing new and previous records; however, these new names are usually listed with their previous synonyms.

At least 41 species of cyanobacteria from 11 genera have been shown to produce geosmin or MIB in the United States. These genera include *Anabaena, Oscillatoria, Phormidium, Lyngbya, Leptolyngbia, Microcoleus, Nostoc, Planktothrix, Pseudanabaena, Hyella and Synechococcus.* These include both planktonic (present in the water column) and attached species.

Algal species differ in size and growth patterns as single cells, clusters or amorphous colonies, coenobia (i.e., in clusters of 2^n (i.e., 2, 4, 8...) cells) and filaments (which can form flakes or large coils). In addition, many colonial cells can break apart during preservation and settling, and these need to be enumerated and included in the total biomass count. Thus, it is important to select the proper magnification for the microscope to view the different algal species, and a working magnification for different size classes is recommended (Figure 2-3). It should be noted that measures of algal cell counts are misleading unless weighted for the substantial differences in size among species.

Size (μm)	Magnification of Microscope
0.2 – 2 (picoplankton)	1000× (oil immersion)
2 – 20 (nanoplankton)	100 – 400×
20-2000 (microplankton, filaments, large colonies)	100×

Sedgewick-Rafter Method. This method has been used more traditionally in the water industry and is relatively inexpensive, requiring the purchase of a compound microscope (preferably of good quality) and specialized chambers. However, unless the sample is very dense or has been preconcentrated, many errors can occur during the determination of an algae count which is based on only 1 mL of sample. The method can generate higher counts and identification errors, particularly when densities are low. Counts can be done

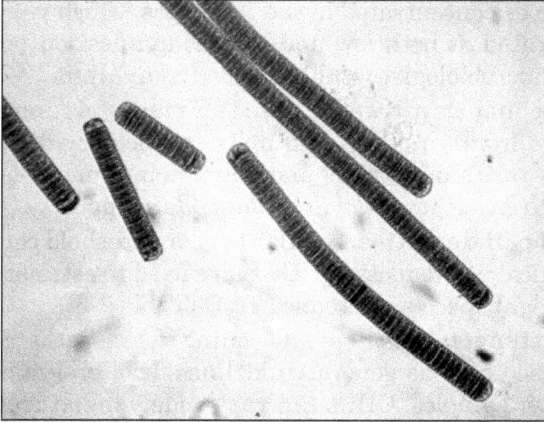

Courtesy of Gary Burlingame.

Figure 2-3 Cyanobacteria as viewed under a microscope

using only 4× and 10× objectives, meaning that small [<20 micrometer (μm)] cells are difficult to count.

This technique uses a 1 mL chamber with a bottom plate divided into 1,000 squares. Before filling with the sample, partially cover the Sedgewick-Rafter chamber with the large rectangular cover slip. Using a pipette, fill the chamber with a well mixed sample, then gently move the cover slip to cover the chamber so that it is snug without any underlying air bubbles; adjust the volume, if necessary, by removing or adding sample. Allow at least 15 minutes for the sample to settle before counting under a compound microscope. Using 10 randomly chosen fields, enumerate every algal cell within the grid field boundaries lying on a boundary line.

$$\text{Algal cells per mL} = \frac{\text{Number of cells counted} \times 1000}{\text{Number of squares observed}}$$

Inverted microscope (Utermöhl) method. This technique is the standard quantitative method for counting phytoplankton worldwide. It requires a good inverted microscope and specialized settling chambers, which can be purchased or constructed to hold specific volumes (e.g., 1 mL, 5 mL) from large cover slips and Plexiglas tube lengths. This method is the most recommended and is widely used across the scientific community, although it is not commonly employed in water utilities. A major advantage of the method is that samples can be concentrated in the chambers, which vary in volume, and enumerated at both low and high magnification power. A limnologist or microbiologist would provide the analysis.

These techniques are well adapted for plankton samples, but benthic samples from scrapings and mats can be very difficult to enumerate as a result of detritus and silt or tough, multi-layered mats (such as with *Lyngbya wollei* and *Cladophora* spp.).

Critical cell densities. Similar to odor threshold concentrations (OTCs), critical cell densities (CDCs) are used to estimate the potential for an algal species to produce T&O (Table 2-6).

These estimates may be misleading for several reasons and should be used only as general guidelines for a preliminary assessment of water samples. CDCs can vary widely in natural water bodies and among different species. Most species produce more than one T&O compound that, when combined, can modify overall odor levels. Finally, the levels and chemistry of T&O compounds produced by a species change with growth and environment.

Table 2-6   Critical Cell Densities for Selected Species of Odor Causing Algae

Algae	Odorant	Cells per mL
Anabaena spp.	geosmin	40 – 200
Aphanizomenon		700
Microcystis spp.	β-cyclocitral	2,000 – 100,000
Oscillatoria		50,000
Oscillatoria cf. chalybea	MIB	100
Oscillatoria tenuis	geosmin	1,000
Phormidium cf. calcicola	geosmin MIB	300 500
Dinobryon spp.	2,4,7-decatrienal 2,4-heptadienal 2,4,7-decatrienal	4,000 50,000 – 400,000 8,000
Mallomonas		500
Synura petersenii	2,6 -nonadienal 2,4,7-decatrienal	300 2,000
Uroglena americana	2,4,7-decatrienal 2,4-heptadienal	100,000 30,000
Ceratium		200
Cryptomonas		1,000
Asterionella formosa	2,4,7-octatriene	3,000
Cyclotella		2,000
Melosira (Aulacoseira)		3,000
Synedra		3,000
Tabellaria		800
Euglena		800
Ankistrodesmus		4,000
Chlamydomonas		4,000
Eudorina (colonies)		80
Pandorina		2,000

Chapter 3

Tracking T&O in Source Water

Diagnosing T&O from Surface Water

If the source water is obtained from a river or open reservoir, it will be necessary to drive around the reservoir or travel upstream of the river to collect samples from different locations. Oil spills or other spills might have occurred upstream. Fire departments may report fires in warehouses storing solvents or other chemicals that could have been washed into the river or watershed by firefighting activity. When sampling in wilderness areas, a global positioning system (GPS) will record the exact location of the sampling points. During upstream sampling events, it is important to use a map showing tributaries and reservoirs as well as wastewater plants and industrial discharge sites.

Because the source of the problem may be located far away, appropriate sampling should be performed by field observations. Common sense is also quite important; for instance:

- If a hydrocarbon odor is detected, look for oil plumes at the water surface. Once located, collect samples from the plume and at upstream locations, as the plume might have travelled a long way. If the plume is located in the middle of the river, sampling from a bridge may be necessary.
- If a chemical-like odor is detected, sampling the different tributaries and industrial areas (upstream and downstream) may help to pinpoint the source. However, chemicals can also be discharged from sewers, and sampling the outlet of wastewater treatment plants may be necessary.
- If a septic or sewage-like odor is detected, discharge points of treated and untreated sewage should be investigated. Specific industries discharging decomposing organic matter, such as paper mills, should be considered.
- If an earthy-musty or fishy, grassy or cucumber-like odor is detected, an algal bloom is the likely cause. A greenish color in the water may indicate such a bloom, although in

some cases algae lie at the bottom of the waterway and are not visible at the water surface. Because an algal bloom can spread over a vast area of the catchment basin, multiple samples must be collected, including from tributaries and lakes or reservoirs. When algae appear in open reservoirs or lakes, a boat will be required to sample areas rich in algal biomass in the water column, and sometimes the use of scuba divers must used to sample attached (benthic) algae.

When monitoring an intake on a surface water source, the following tests are useful: algae counts and types, chlorophyll-a, pH, water temperature, turbidity, TDS, conductivity, total and dissolved iron, total and dissolved manganese, geosmin and MIB, and VOCs.

Diagnosing T&O from Algae and Cyanobacteria

Algae can be the source of a variety of VOCs released during their growth, cell lysis or death, or when the microbiological cells decay. The odors produced can be nitrogen-based (e.g., trimethylamine and propylamine), emitting such odors as ammoniacal and fishy. Odors can be sulfur-based (e.g., dimethyl-sulfide and dimethyl-trisulfide), emitting such odors as septic, pungent, and rotten eggs. Terpenes and terpenoids (e.g., geosmin, MIB, and β-cyclocitral) emit odors such as musty, earthy, tobacco-like, violet-like, and floral. Hydrocarbons (e.g., 2,4-decadienal, n-hexanal, 2,6-nonadienal, octan-1-ol and pent-1-en-3-one) emit such odors as fishy, rancid, green vegetation, grassy, cucumber-like, earthy, geranium-like, and fruity.

Whether the chemicals produced by algae will affect the odor quality of the water depends on many factors, including but not limited to the type of algae and their density, where the algae grow, the water quality conditions surrounding them, and the conditions of the waterway where they are growing.

Phytoplankton are important because of their photosynthetic activity; they are primary producers in the food chain. They can also serve as a water quality indicator. Procedures for sampling for phytoplankton at a treatment plant intake often involve dip or grab sampling from the surface or near surface of the water. Dip sampling captures the phytoplankton or microscopic flora that float. Sampling does not typically capture the periphyton, or microorganisms attached to submerged natural or artificial substrates, nor does dip sampling capture the benthic algae that prefer to live on the bottom of lakes and rivers.

Planktonic algae can become significant when they bloom. Algae blooms are a visible abundance of algae. Three scenarios can exist in a river watershed: algae blooms can occur in upstream reservoirs and be discharged to the river and continue to grow as they travel down the river; algae blooms can occur in the river itself; algae can accumulate in quiet reaches and pools during low, steady flows and be washed out when flows increase, such as during periodic reservoir releases.

Reservoirs are also tricky to sample. Algae can occur at various depths and along the shoreline or on the bottom in benthic mats. Thermal stratification may encourage reservoir blooms, and different types of algae may be found at different depths. Finally, algae blooms may occur in specific coves or bays of a reservoir rather than throughout the entire reservoir.

In one case, *trans*-2, *cis*-6-nonadienal (a cucumber odorant) in drinking water was found to come from a reservoir in the source watershed during winter periods when the reservoir was iced over and topping its spillway. The growth of algal flagellates under the ice produced the odor. When the odor occurred in the treatment plant's finished water over 200 miles downstream, algae could not be correlated to the odor because the algae did not travel downstream from the reservoir, even though the odor chemical produced by the algae did travel downstream.

In another case, when a water treatment plant experienced a geosmin odor, the results of routine dip samples collected from the plant intake during the morning were evaluated to determine which algae might be causing the problem. A floating bloom of the diatom *Cyclotella* was found. However, relevant literature had never reported any geosmin-producing diatom.

Upon further investigation, a bloom of *Oscillatoria* was discovered in benthic mats on the river bottom. Grab sampling in the morning did not detect it. Sampling was changed to the afternoon, when portions of the benthic mats broke from the bottom and floated to the water surface and then downstream by the treatment plant intake. However, even in this case, a sample collector might avoid collecting samples from the scum on the surface of the river, not realizing that it was the cyanobacterium of concern.

Adding to this complexity is the possibility that the odor may not be present in the source water containing the algae. The odor may be released by the algae during treatment, such as when the algal cells are lysed open by chlorination and oxidation of the algae. In addition, their extracellular or intracellular materials may produce byproducts that contribute odors.

For example, the chlorination of algae can produce aldehydes that impart a grassy, vegetative odor to water. Algae that die en masse can collect in sedimentation basins and on filters and decay, where the proteins and amino acids form organic sulfides (i.e., dimethyltrisulfide) that impart marshy, sulfurous odors to water. When an algal bloom occurs and an "off" odor occurs in the drinking water, it may be important to sample throughout the treatment process to determine whether the odor is formed during treatment.

Strategic plans must be developed when studying algae to determine the source of a T&O problem and where the algae are growing. Sample collection methods and sites must be carefully selected. Sample collection might involve sampling at different depths of water and at different times of day. Sampling might also require collection before and after treatment. For example, *Chrysosphaerella* spp. (a pigmented flagellate that can be part of the phytoplankton along with algae) has been associated with a metallic flavor that occurs after the organism has been exposed to chlorine.

Diagnosing T&O from Spills and Discharges

During an incident that occurred on the Ohio River in 1989, discharge from a resin manufacturing plant was transported to a sewage treatment plant, which in turn discharged into the Ohio River. Chemical odors caused by dioxolane derivatives were detected in drinking water supplies as far as 137 miles downstream from the point of wastewater discharge.

Sampling points were investigated during another incident caused by dioxanes and dioxolanes from a polyester resin manufacturing plant (indicated as A in Figure 3-1) north of Barcelona (Spain). River water samples (circled) both upstream and downstream of the resin manufacturing plant, as well as numerous groundwater samples (numbered), were collected and analyzed during this incident.

Spills into rivers may contaminate underlying aquifers, and underlying aquifers can also contaminate surface water. During one event, an ether-like odor in drinking water was traced to groundwater contaminated by ethyl *tert*-butyl ether (ETBE). The groundwater had contaminated the nearby surface water during a rise in the water table.

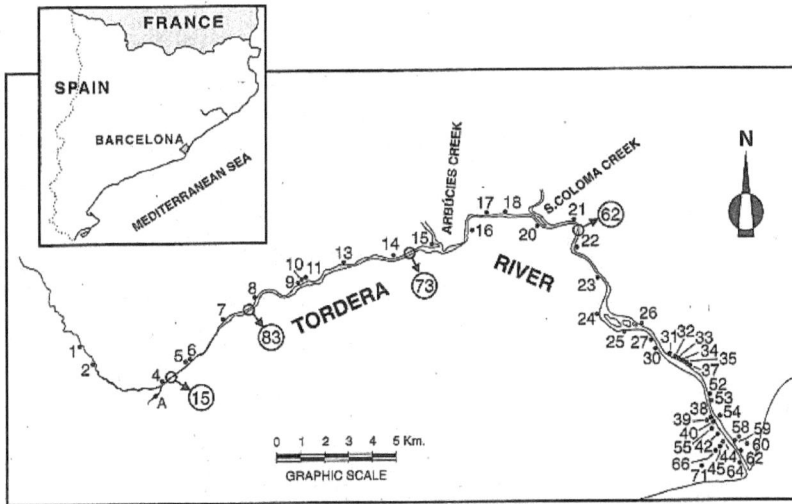

Study area and sampling site locations in the Tordera aquifer. Groundwater samples are numbered, and river water samples are encircled. Wells 4–8, 13, 15–18, 24, and 30 are used by industries; wells 23, 42, 45, 54, 60, 62, 66, and 71 are used for agricultural purposes; wells 14 and 64 are piezometers, and the rest are used for drinking water purposes.

Figure 3-1 Example of sample site selection in Spain to track a chemical discharge (*Environmental Science and Technology*, 1998.)

Diagnosing T&O from Groundwater

Aesthetic issues in groundwater are often related to one of four water quality constituents: iron, manganese, ammonia, and hydrogen sulfide. These constituents can occur naturally in the groundwater as the result of microbial activity in the well.

Iron and sulfur bacteria are problematic. Odors related to sulfur are often detected by the human nose more effectively than relying on laboratory analysis, because the human nose is very sensitive to these compounds. In addition, these compounds tend to change when in contact with air or volatilize during sampling, resulting in unrepresentative laboratory results.

Basic tests can be used to help monitor the quality of a groundwater supply. These tests include hardness, pH, dissolved oxygen, conductivity, total and dissolved iron, total and dissolved manganese, hydrogen sulfide, turbidity and VOCs.

Chapter 4

T&O Control and Prevention Options

Typical Treatments for T&O Control

Treatment processes to control T&O include those that oxidize the compounds causing the taste and odor converting these compounds to another form or to completely mineralize them. Other processes remove the compound(s) from the water or cause them to volatilize, which removes them from the water. The types of treatment processes used for T&O control include aeration, oxidation, adsorption, removal, and biodegradation.

Aeration is used effectively to treat VOCs and hydrogen sulfide, so it is often used in groundwater treatment. Aeration is rarely effective in the treatment process of surface water.

Groundwater Options

Groundwater problems include chemical contamination, salt-water intrusion, high mineral content, iron-reducing bacteria, and sulfur bacteria. A groundwater well may be abandoned because it would be impossible or too costly to treat the water and manage its quality.

Rotten egg-smelling water and a black color may indicate sulfur bacteria, which can be present in shallow and deep wells. Sulfur odor detected in wells can be treated by oxidizing the water with potassium permanganate (such as 4 mg/L $KMnO_4$ per mg/L of H_2S) or with chlorine (such as 2 mg/L chlorine per mg/L H_2S). Organically bound sulfides are more difficult to control, especially with chemical oxidation. A stronger oxidant such as ozone may be required, or longer contact times (up to 25 minutes) may be required for sufficient oxidation to occur.

Other treatment alternatives include manganese greensand media filtration, which is effective for iron and manganese as well as sulfides (with adequate upstream oxidation), and granular activated carbon (GAC) specifically designed to catalytically oxidize sulfides. Sufficient contact times are needed for these oxidation processes to work.

Elevated levels of iron may indicate the presence of iron bacteria in the well. It is highly desirable to eliminate the iron bacteria rather than to rely on treatment of the well water to remove the iron. If the water tests positive for iron bacteria, rehabilitation and repair of the well or well relocation may be recommended. Shock chlorination, acidification, and physical scrubbing may remove the problem bacteria. Aeration and filtration may be needed to control the iron in the water if its production cannot be mitigated within the well.

Chlorine

Chlorine is the single, most common chemical used for both disinfection and oxidation of drinking water. It can be purchased as a gas and diffused into the water or purchased as a hypochlorite solution. Existing risk management regulations and new security regulations discourage utilities from storing large quantities of gaseous chlorine at treatment plants, so on-site generation of chlorine has emerged as a third alternative. Chlorine works in two ways to help control taste and odor: direct oxidation of odorants and control of odor-producing microorganisms.

Specific chlorine species include (in order of decreasing oxidizing strength) hypochlorous acid (HOCl), hypochlorite (OCl⁻), and monochloramine.

Chlorine is effective for oxidizing the chemical species that cause fishy, sulfurous, cucumber, grassy, decaying vegetation, marshy, and septic odors. Chlorine is also effective for controlling the growth of algae within a treatment plant; therefore it can help control earthy/musty odors, such as those caused by MIB and geosmin. However, chlorine will also lyse algal cells, releasing the intracellular contents to the water. Chlorine can also be used to oxidize iron and manganese.

Potassium Permanganate

Potassium permanganate ($KMnO_4$) is typically used as a preoxidant at the head of the treatment plant. It is often preferred to chlorine, which forms regulated DBPs such as trihalomethanes and haloacetic acids. Preoxidation with potassium permanganate reduces the chlorine demand of the water so that a lower dose of chlorine can be used downstream to maintain a given residual in the finished water of the plant. Preoxidation also helps with iron and manganese control.

KMnO$_4$ is effective for reacting with compounds containing a double bond between carbon atoms (C=C double bonds) and functional groups containing oxygen such as alcohols, phenols, and alkylbenzenes. Potassium permanganate is effective for controlling odors such as fishy, grassy, and cucumber. Typical potassium permanganate dosages range from less than 0.5 mg/L to 3 mg/L.

Care must be taken to avoid overfeeding potassium permanganate because the residual permanganate produces pink-colored water. Because the dosage of potassium permanganate can be limited, and it is not effective for all tastes and odors, it is often used in combination with other chemicals. For example, it is frequently added upstream of PAC or chlorine.

Chlorine Dioxide

Chlorine dioxide does not produce DBPs, such as trihalomethanes (THMs), and it prevents the formation of chlorophenols, which are produced when water containing phenols is chlorinated. Chlorine dioxide is not recommended for the control of alcohols and aldehydes, and it is not effective for controlling the cucumber odor produced by *trans*,2-*cis*,6-nonadienal. It can be useful for iron and manganese control, especially when the water has a significant chlorine demand.

The dosage of chlorine dioxide rarely exceeds 2 to 3 mg/L because it degrades into the regulated by-product chlorite, which has a USEPA maximum contaminant level (MCL) of 1 mg/L. Roughly 60 percent of the dosed chlorine dioxide ends up as chlorite in the finished drinking water.

Ozone

A powerful oxidant and disinfectant, ozone is used to oxidize a wide range of constituents in water, including iron and manganese; to remove color; and to address a variety of odorants, including MIB and geosmin (Figure 4-1). Because it does not form chlorinated DBPs, it may be used instead of chlorine. Ozone in combination with downstream chlorine is effective for controlling fishy, decaying vegetation, septic, and sulfurous odors.

Ozone can follow two general pathways: direct oxidation by molecular ozone and indirect oxidation by forming hydroxyl radicals. Direct oxidation by ozone is highly selective but exhibits slow kinetics (takes time to fully react). Indirect oxidation by the hydroxyl radical is fast and nonselective and is therefore preferred for treating odors caused by a mix of different chemicals.

When disinfecting with ozone, a measurable residual of molecular ozone must be maintained for some time to obtain disinfection credit. Direct molecular oxidation is favored under conditions of lower pH (less than 8), moderate to high alkalinity, and modest levels of organic carbon. Under conditions of high pH (greater than 8 to 9), ozone reacts rapidly. Under these conditions, it is more difficult to maintain measurable ozone residual to achieve disinfection credit.

Excessive ozone dosages can result in fruity and sweet odors, whereas low ozone dosages may not be sufficient to oxidize the odorants in the source water.

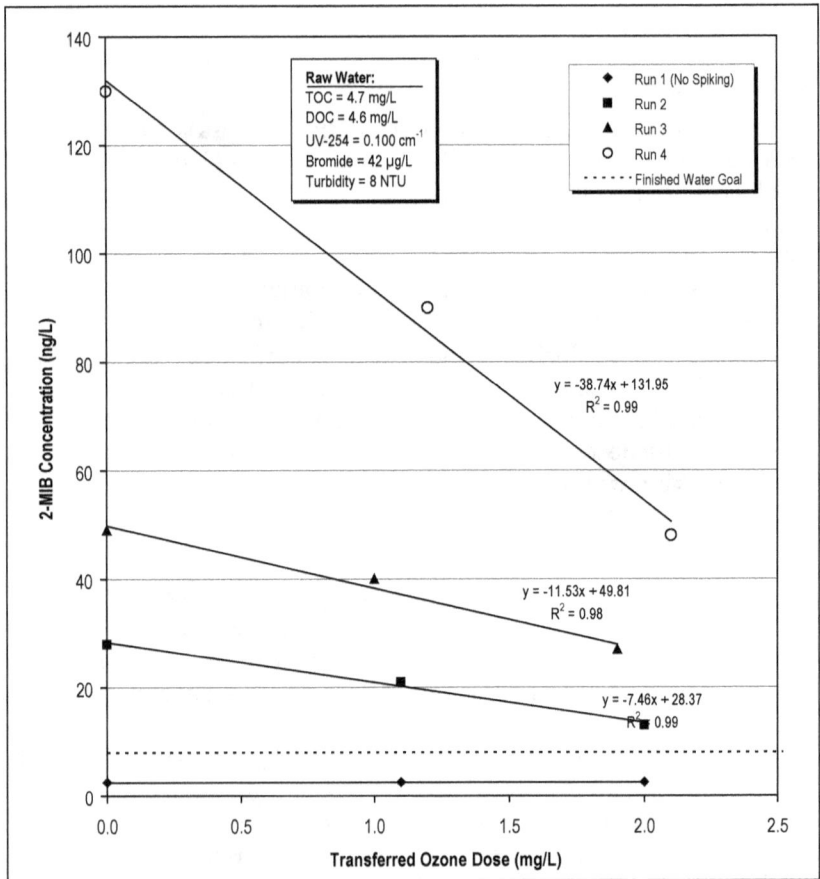

Courtesy of AWWA 2002 WQTC.

Figure 4-1 Example of measured removal of MIB by ozonation at three different spiked concentrations

Ultraviolet Light Irradiation

Ultraviolet light (UV) irradiation is an alternative to oxidation and disinfection that does not form regulated DBPs. UV irradiation has gained interest as a control technology for the disinfection of pathogens such as *Cryptosporidium*. It is typically used along with and after other treatment processes; in the United States, UV is followed by the addition of a disinfectant residual for water distribution.

UV photolysis for the oxidation of organic chemicals in water has been the subject of research studies. The current understanding is that at typical doses for disinfection of drinking water (approximately 40 mJ/cm^2), there is almost no chemical degradation of geosmin or MIB. It appears that the UV dose would have to be closer to 500 mJ/cm^2, which is more than ten times the typical dose used for disinfection, to achieve about 90 percent destruction of T&O chemicals. However, the addition of hydrogen peroxide to water in the presence of UV light shows promise for obtaining sufficient T&O control.

Advanced Oxidation Processes

Advanced oxidation processes (AOPs) exploit the use of the hydroxyl radical to nonselectively and rapidly oxidize odorants. Examples include:

- Ozone with hydrogen peroxide
- Ozone with UV radiation
- Ozone at elevated pH (8–9)

Advantages of AOPs are that they can be effective at a lower ozone dose compared to oxidation with molecular ozone; they are less dependent on contact time because free radicals react very quickly; and they effectively treat many odorants and other regulated chemicals, such as volatile and synthetic organic chemicals, pesticides, and herbicides.

Activated Carbon

Activated carbon can achieve good control of geosmin and MIB (Figure 4-2). Activated carbon has the additional benefits of not forming DBPs as well as removing precursors and other contaminants. GAC is used either as a filter medium in a combined filtration/adsorption process or in a post-treatment contactor. PAC can be applied at various locations in the treatment train or on top of the filters as a filter cap. GAC is generally preferred over PAC if odor problems occur frequently and are sufficiently serious that very high dosages of PAC would be required.

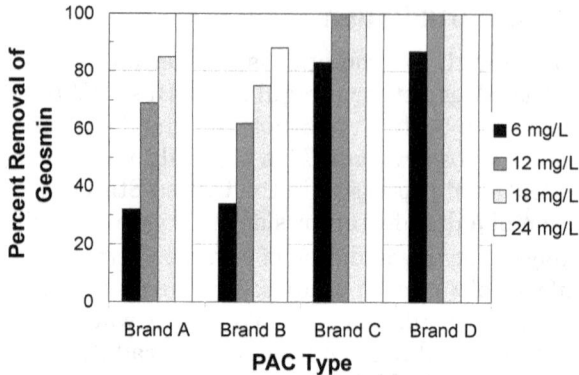

Figure 4-2 Example of jar testing data for the comparison of geosmin removal (spiked at 15-25 ng geosmin/L) using four different brands of PAC

Typical PAC dosages are as follows:

Typical routine dosage:	2–8 mg/L
Intermittent use in episodes:	5–20 mg/L
Emergency use:	20–100 mg/L

Variability in the levels of natural organic matter can be problematic because organics compete for adsorption sites on the carbon. In addition, the presence of oxidants (chlorine, ozone, permanganate, or chloramine) can produce surface oxides on the carbon that reduce its adsorption effectiveness.

Considering the generation of solid residuals at a treatment plant and typical PAC feeding equipment, a PAC dosage of 50 mg/L is a practical upper limit for PAC use for any significant length of time. In some surface waters, the routine use of PAC has been shown to also reduce DBP precursors.

Ozone followed by downstream GAC can be used to control ozone-produced aldehydes and ketones (sweet, fruity odors) because GAC develops a biofilm that biologically removes these odorants. Recent research has shown that such biological filtration processes can effectively remove geosmin and MIB at low levels (5 to 50 ng/L).

Both PAC and GAC remove free chlorine residuals catalytically and rapidly, so the addition of chlorine should preferentially not come in contact with the carbon. Preoxidation of the water with potassium permanganate seems to improve PAC effectiveness. PAC removal efficiency in the same water source and under the same conditions will stay the same at all odorant concentrations. Therefore, a given dosage of PAC is expected to result in the same percentage removal of MIB, regardless of the concentration of MIB in the untreated water.

Various types of carbon are available on the market. Carbon brands can vary from less than 70 percent to more than 50 percent geosmin and MIB removal efficiency. Different carbons are manufactured from different raw materials, including coal, wood, and coconut shells, and different activation processes are used to increase the surface area of the carbon to enable effective treatment. Activation is often performed using high temperature and pressure steam or liquid acid. No single brand of carbon meets all needs.

Bench-scale tests should be performed on several available carbon types to test for their effectiveness on a site-specific basis. Different carbons perform differently, and a lower dosage of one carbon may perform as well as a higher dosage of another. The dosage required to meet treatment goals should be considered when comparing carbon costs and preparing carbon specifications. The carbon with the lowest cost per pound may, in fact, be more expensive if a higher dosage is required to achieve treatment goals.

Biologically Active Filtration

Biologically active filtration includes carbon, sand, and mixed media filters. In general, biological growth is encouraged in this type of filtration. For example, ozone may be added before the filters to provide dissolved oxygen for aerobic microorganisms. Ozonation may also reduce complex natural organic matter into more readily available organic carbon for microorganisms to consume, thus supporting their growth on the filter media. Chlorine, which stops the biological growth, is not added until after the filters.

These filters should be designed carefully based on the characteristics of the incoming water. Their effectiveness in removing various T&O compounds varies widely, and upfront pilot studies should be conducted. These filters can be effective for geosmin and MIB removal.

Membranes

The last decade has seen major improvements in membrane technology, although the removal of T&O compounds is not the main impetus for the application of membranes in drinking water treatment. Membrane processes that have applications in drinking water treatment include microfiltration (MF), ultrafiltration (UF), nanofiltration (NF), reverse osmosis (RO), and electrodialysisreversal (EDR). In general, factors that affect the selection of a membrane application are pore size, surface chemistry, flux, recoveries, and degree of fouling.

Membranes are classified based on the size of the compounds that they can filter and the molecular weight cutoff (MWCO) (Figure 4-3). Membrane pore size is the main parameter controlling the use of membranes for a particular application.

T&O compounds, such as geosmin and MIB, are found in dissolved form (algae extracellular products) in surface water or in bound form within algal cells (intracellular). Intracellular algal metabolites may be released on breaking or lysis of the algal cells, which can add high levels of T&O compounds to the water. In general, low pressure membranes (MF and UF) cannot remove extracellular algal metabolites of small molecular weight. However, research has shown that they can be used successfully to remove cyanobacterial and algal cells (1 micron or larger) and intracellular T&O compounds.

Under normal operating conditions, MF (~ 0.3 µm pore size) and UF (~ 0.01 µm pore size) membranes can effectively remove cyanobacterial and algal cells with minimum damage to the cells. Low-pressure membranes used alone cannot remove algal metabolites. However, they can be effective in combination with other membranes (known as *integrated membrane systems*) or other conventional treatment processes.

One study found that although UF is not expected to remove T&O compounds in treated drinking water, its combination with PAC

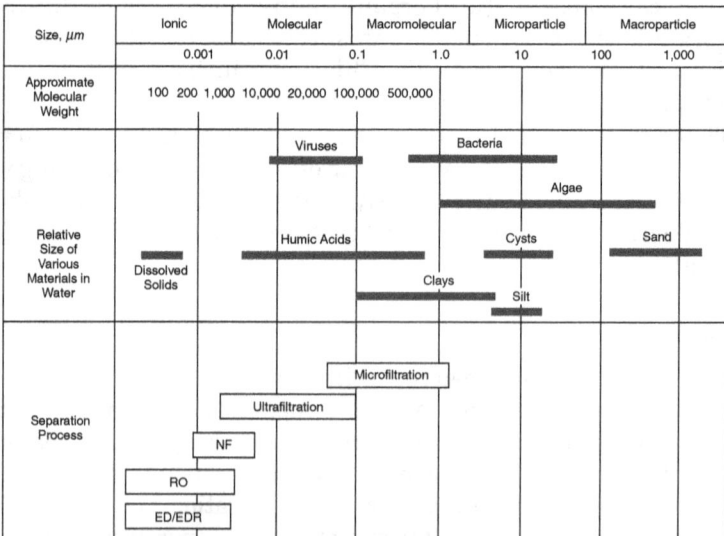

Figure 4-3 The efficiency of membranes based on pore size and molecular weight cutoff (MWCO)

can effectively control T&O. A simulation of the UF–PAC combination demonstrated that the process could handle prolonged episodes of geosmin concentrations up to 50 ng/L.

The removal of geosmin and MIB by high-pressure membranes (NF and RO) has been the subject of several studies. These types of membranes, with the appropriate size ranges, can remove a range of extracellular dissolved algal metabolites.

In general, the studies reported removals that are highly dependent on the type of membranes used. In one study, researchers observed that cellulose acetate membranes removed 35 to 50 percent of MIB and geosmin, while polyamide membranes achieved up to 99 percent removal. In the case of low molecular weight T&O compounds, such as dioxanes and dioxolanes, removal efficiencies in the range of 74 to 88 percent were reported when NF was used. One study observed NF and RO processes to efficiently reduce 2,4,6-trichloroanisole (TCA), one of the most odorous compounds known, as well as its precursor, 2,4,6-trichlorophenol (TCP), to below their odor detection thresholds.

Reverse Osmosis and Desalination

The USEPA SMCL for TDS is 500 mg/L. TDS is a measure of the total ion concentrations, including cations such as calcium, magnesium, potassium, sodium, aluminum, iron, manganese, and anions such as bicarbonate, carbonate, chloride, sulfate, and nitrate. TDS varies significantly depending on the water source and local geology. Reverse osmosis treated water and distilled water can have TDS values less than 10 mg/L.

Desalinization using RO treatment, which is designed to remove ions, can result in significantly lower level of minerals in treated water. Specifically, RO removes calcium and magnesium along with carbonates, but sulfate, chloride, and TDS can still pass through in detectable quantities.

RO treatment can have a noticeable impact on tap water taste because the loss of minerals can leave a slick feeling and the loss of alkalinity can leave a drying, bitter taste. Decreased alkalinity also leaves the water more corrosive. Corrosive water in turn increases the leaching of metals from water pipe and plumbing, which can impact the taste of the tap water.

Because desalination removes 95 to 99 percent of minerals and can reduce pH, it is typically followed by remineralization and restoration of alkalinity. Remineralization consists of the addition of lime, which contains calcium carbonate and some magnesium carbonate.

Caustic soda, bicarbonate, sodium carbonate, phosphates or silicates can also be used to restore minerals to the water.

Nontreatment Options

The best control strategy is prevention, where possible. This includes preventing T&O-causing chemicals from being produced or introduced into the water, or avoiding the use of the tainted water altogether. Industrial pollution controls, minimizing spills to waterways, and upgrading wastewater treatment are clearly sound, preventive options.

Watershed management is part of a water utility's source water protection program. This includes nutrient control (particularly nitrate, ammonia nitrogen, and orthophosphate) from point and nonpoint discharges to a watershed. Such nutrients encourage algal blooms and aquatic vegetation growth.

In reservoirs, methods can be used to alter water temperature, light availability, water clarity, water depth, and water flow, all of which can alter microbiological populations and preselect for less annoying algal species. Reservoir controls include artificial destratification and hypolimnetic aeration, which alter the oxygen and temperature layers of the water to affect the algae that may grow in those layers. Limnology is the study of the ecology of a water body; limnologists can provide important information for better reservoir management.

Copper sulfate is the most common algicide used, typically in reservoirs. It comes in various crystal sizes, from a fine "snow" that dissolves quickly to very large crystals that sink to the bottom, depending on where the problem algae are growing. Algae vary in their sensitivity to copper sulfate and algicides. Some algae are fairly resistant. It is possible that, while treating to eradicate one problem, conditions that encourage a more resistant, more annoying algal species are enhanced. In addition, applications that kill algae may cause the release of intracellular compounds, including toxins.

If an algal bloom is underway, it is important to minimize the intake of algal cells into the treatment plant. Cell lysis, such as during chlorination, can release high levels of odor-causing chemicals to the water. Coagulant aids and pretreatment can help reduce the carryover of algae.

The treatment plant or water supplier may have the option of avoiding water containing T&O problems or of blending it with other water. Perhaps a tainted well can be shut down; a river intake can be closed until the river's flow has pushed the problem past; or the reservoir intake can be drawn from a different depth where water quality is more acceptable.

Chapter 5

Making Treatment Decisions

Making Decisions Using Customer Complaint Data

Customer complaints provide valuable feedback concerning changes to finished water quality that can be used to identify T&O problems. Many water quality monitoring programs take customer feedback very seriously because customers are distributed throughout the system, sample the water at all hours, and can detect slight changes in water taste, odor, and appearance. Customer feedback is becoming increasingly necessary as water utilities strive to improve water quality and customer relations. For example, taste, odor, color, and clarity complaints can signify the presence of contaminated drinking water. A customer complaint response protocol will save time and reduce the number of people exposed to contaminated water.

Complaints related to T&O must be separated from other types of complaints and gathered in one place (Figure 5-1). A utility should centralize the contact location for receiving customer input and sort complaints into broad categories. Water quality complaints can then be further categorized. T&O complaints should comprise a separate category and exclude other common complaints such as "no water" or "flow problems." Finally, the taste or odor complaint should be specifically identified using common and consistent descriptors to identify the taste or odor of concern.

Consistent terminology should be used to describe T&O complaints. A standard check-sheet can be used by everyone processing customer complaints (Figure 5-2). It is not acceptable for different complaint managers to use different terms for similar odors, such as sulfur, eggy, rotten eggs, H_2S, and rotten. Using different terms and different spellings of the same term leads to confusion and becomes impossible to sort electronically. Consistent terminology allows customer complaints to be categorized and tracked properly over the short and long term.

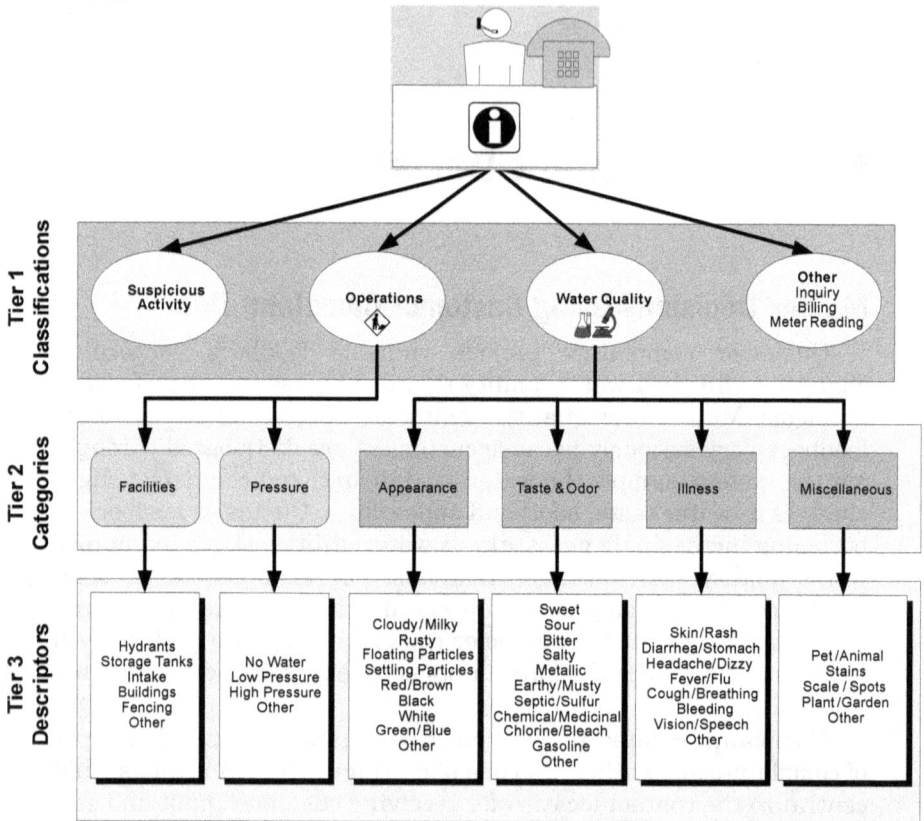

Figure 5-1 The three-tiered approach to harnessing customer complaints

Once properly sorted and categorized, all T&O complaints must be reviewed to determine the number and types of complaints that occur. If many customers call to report similar or dissimilar problems, a potential problem may exist, and the cause should be investigated. An understanding of the "typical" conditions for the customer base needs to be developed so that they may better detect changes in water quality. The analysis of complaints in an historical context allows a utility to better track water quality at the customers' taps.

Frequency analysis based on total water quality complaints is relatively easy to implement (Figure 5-3). It is simply a count of the number of complaints during a specific time period (e.g., the number of complaints per day). For larger utilities, specific categories of

Proposed Customer Feedback Check Sheet

PWSID No. 000000000   Water Plant #: (000) 000 - 0000
                                       Customer Service #: (000) 000 - 0000

RECEIVING INFORMATION		
Customer Name:	Date:	Follow-Up Needed? Yes / No
Address:		Go to Work Order
Telephone:	Time:	Department Notified? Yes / No

CODING		
OPERATIONS	**WATER QUALITY**	**WATER QUALITY**
FACILITY ISSUES	*TASTE AND ODOR ISSUES*	*MISCELLANEOUS ISSUES*
√ Descriptor	√ Descriptor	√ Descriptor
Hydrant	Sweet	Pet / animal
Storage Tank	Sour	Stains
Intake	Bitter	Scale / spots
Building	Salty	Plant /Graden
Fencing	Metallic	Other
Other	Earthy/Musty	
	Septic / Sulfur	**SUSPICIOUS ACTIVITY**
PRESSURE ISSUES	Chemical / Medicinal	
√ Descriptor	Chlorine / Bleach	
No Water	Gasoline	
Low Pressure	Other	
High Pressure		
Other	*ILLNESS ISSUES*	**COMMENTS**
	√ Descriptor	
WATER QUALITY	Skin / Rash	
APPEARANCE ISSUES	Diarrhea/ Stomachache	
√ Descriptor	Headache / Dizzy	
Cloudy / Milky	Fever / Flu	
Rusty	Cough / Breathing	
Floating particles	Bleeding	
Settling particles	Vision / Speech	
Red / Brown	Other	
Black		
White		
Green / Blue		
Other		*(additional description space should be available on back side of paper)*

Water Investigation ID No. 000000000

Courtesy of Journal AWWA.

Figure 5-2 Check sheet for recording water quality complaints

complaints can be considered in addition to the overall number. For frequency analysis to be useful, a historical basis is needed to compare normal versus abnormal. A mean complaint frequency might be 0.11 to 2.64 complaints per day. Adjusted to a customer base, the mean frequency might become 0.2 to 2.60 complaints per 1,000 customers per year. The most common temporal pattern is a notable lower frequency on weekends because most utilities' call centers are closed on weekends; thus fewer calls are recorded. Utilities with multi-year records average 8 to 9 times the average number of water quality complaints on a weekday as a weekend day.

Short time periods (i.e., one day), are useful for detecting sudden changes in complaint counts. Longer time periods are useful for detecting gradually worsening problems (Figure 5-3). For example, a utility with a single day alarm of six complaints per day would not begin an investigation based on four complaints. But if four complaints were made for 4 or 5 days within a 7-day

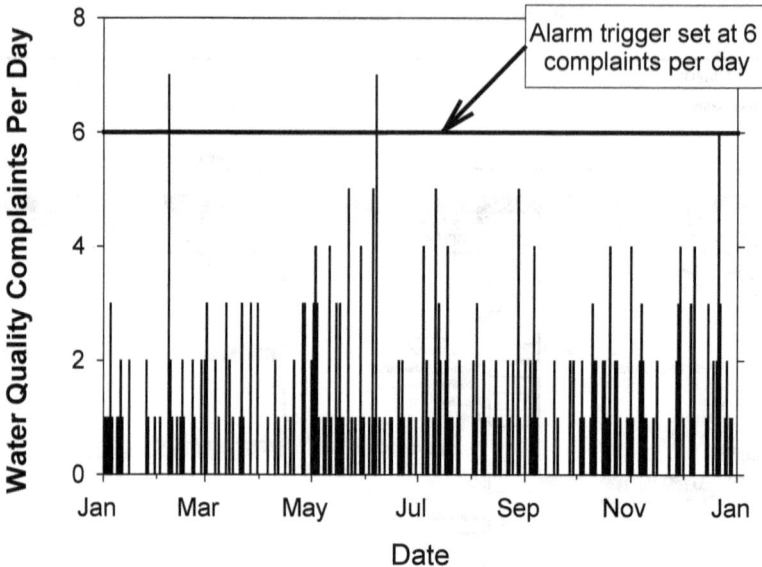

Courtesy of Dan Gallagher.

Figure 5-3 Example of a control chart for customer complaints

period, this would likely constitute an unusual event that should be investigated.

In summary, for effective use of customer complaints

- Begin collecting and maintaining complaint data, and continue to do so.
- Learn what is typical for the utility. Evaluate seasonal trends, weekday/weekend trends, typical numbers of complaints, and typical descriptors used by your customers.
- Review the data daily as part of a standard operating procedure.

If the cause of the T&O problem is known, such as with geosmin, it is possible to correlate the complaints to the concentrations of geosmin (Table 5-1). This helps to establish treatment goals and early warning triggers. However, the collection of customer complaint data must result in good quality information.

Making Decisions Using Treatment Data

Gathering good data to document the effects of treatment of a T&O problem can be difficult when the goal of treatment is to eliminate

Table 5-1 Correlation of customer complaints to the concentration of geosmin during a geosmin earthy odor episode

Concentration of Geosmin in Drinking Water (ng/L)	Number of Customer Complaints Received in a Day	Concentration of Geosmin in Drinking Water (ng/L)	Number of Customer Complaints Received in a Day
88	48	39	2
77	22	36	4
67	8	28	1
50	12	24	3
48	15	20	2
46	10	19	0
43	5	16	0
42	7	9	0
41	6		

the problem rather than to study it. It is preferable to set treatment and hold it until a steady state in treatment is reached. However, treatment constantly changes for factors such as coagulation and chlorine addition and pH control. PAC or $KMnO_4$ or chlorine dioxide may be added or adjusted based on the T&O data, resulting in too many variables and too little data. The T&O problem ends, leaving a chaotic mess of tangled, undecipherable data.

Ideally, some treatment facilities have pilot plants or side treatment trains where they can run trials with treatment changes and study the impacts on T&O in controlled experiments. A jar testing apparatus can be used to conduct some experiments, but odor tests conducted on treated water from jar testing can be difficult.

Operations data for the treatment plant can be reviewed to provide insight to help identify the cause of T&O problems. Changes in coagulant and oxidant demands should be noted, as they may indicate changing water quality and the increased presence of reduced chemical compounds in the supply source. If an algal bloom is developing in the supply, diurnal pH swings may reflect the photosynthetic activity of the algae. Such pH variations can be more easily detected than the algae themselves and require adjustment of other treatment chemicals to maintain adequate treatment efficiency.

Regular visual inspections that are often used to monitor treatment processes can also provide useful information for identifying the cause of T&O problems. Visual inspections can help identify the presence of algae in the source water or within treatment process basins, color changes in the water or sludge, and slimes or stains on treatment processes. More thorough visual inspections can be performed during shutdowns when basins, filters and other facilities are drained, exposing areas difficult to see when the plant is in operation.

When monitoring treatment processes for changes that affect T&O, the following should be considered: chlorine and/or ozone demand, coagulant demand, ammonia-nitrogen, total chlorine residual, dichloramine residual, pH, turbidity, and alkalinity.

Sampling Through Treatment Processes

After a utility determines that a T&O problem may be present in the water leaving the treatment plant, it is advisable to collect and evaluate samples throughout the treatment process. It often takes time to mobilize an intensive sampling and analysis effort, and it is easy to miss the peak of the problem, resulting in analysis of unrepresentative samples containing little or no odor. Utilities are encouraged to maintain an ongoing sampling program so that a mobilization effort is not required in the midst of a T&O crisis.

Most T&O events that need to be evaluated through a treatment process are caused by chemicals that impart an odor to the water rather than a taste. Thus, testing of partially treated water should be limited to odor testing or chemical analyses to prevent exposure to health hazards. Additional care should also be taken when high concentrations of hazardous compounds could cause health concerns. Before performing sensory analysis, suspect samples should be chemically analyzed to determine whether hazardous chemicals are present. If solids are present, such as flocculants and coagulants, they should be settled out before conducting odor analysis.

Samples with high levels of chlorine should be dechlorinated because chlorine can mask odors in the samples. Hydrogen peroxide and ascorbic acid can be used to successfully reduce chlorine (Tables 5-2 and 5-3). Hydrogen peroxide is not efficient for reducing chloramine because the kinetics are too slow. Overdosing these chemicals could also affect the odor of the sample. Hydrogen peroxide works well to remove the chlorinous odor of chlorine. When used for the removal of chlorine and chloramines, sodium sulfate can contribute a slight "concrete" odor.

Samples should be collected well after ozone treatment to ensure that the ozone residual is not present during odor analysis.

Various methods can be used to conduct sampling. With grab samples, a sample bottle is lowered into the water or filled at a tap. Samples should be collected from precleaned glass bottles and used only once to prevent the carryover of contaminants from one sample to another. It is preferable to have pressurized taps installed that provide representative samples from the process flow.

Anticipating a T&O Problem Using Early Warning

It is preferable to have an early warning system or predictor of problems in place. This is especially helpful for minimizing customer complaints and news media attention. An early warning system can directly monitor T&O compounds, such as TDS, geosmin, or hydrogen sulfide. In addition, an early warning system can monitor indicators of conditions conducive to T&O, such as pH, dissolved oxygen, chlorine demand, and algae types and numbers. An early warning system may also monitor general environmental conditions such as drought, flood, or forest fire, or changes in treatment that lead to the presence of T&O problems.

Some utilities may not have the capability to continuously monitor chemicals that cause T&O in water. Some analyses are expensive or require a contract with a private laboratory. Other analytical methods are not yet standardized. Nonetheless, by gaining an understanding of the sources and causes of T&O problems, a utility can establish early warning procedures specific to its source water and treatment.

Table 5-2   Dosages of Dechlorinating Agents

Dechlorination Agent	Effective dosage (mg agent/mg chlorine as Cl_2)
Ascorbic acid	3.0
Hydrogen peroxide	0.6

Courtesy of Standard Methods for Examination of Water & Wastewater.

Table 5-3   Dechlorinating Agent Dosage Chart for Chlorine

Agent	Strength of Dechlorinating Agent	Approximate Total Chlorine Concentration (mg/L as Cl_2)	Dosage of Dechlorinating Agent
Ascorbic Acid	5 g/L	1	1.2 mL
		2	2.4 mL
		3	3.6 mL
		4	4.8 mL
Hydrogen Peroxide	3% (drug-store grade)	1	40 µL
		2	80 µL
		3	120 µL
		4	160 µL

Courtesy of Standard Methods for Examination of Water & Wastewater.

In one case, years of monitoring algae and geosmin in the river revealed that geosmin (an earthy odor produced by cyanobacteria) occurred seasonally when benthic (bottom-dwelling) cyanobacteria (blue-green algae) bloomed. A specific pattern of rainfall and spikes in river flow helped to flush the bottom dwellers down the river. A routine monitoring program was subsequently established based on season, river flow, and geosmin analyses.

In another case, *trans*-2, *cis*-6-nonadienal (which emits a cucumber odor) occurred when a reservoir over 200 miles upriver topped off. Although testing for the chemical at low levels was difficult, the conditions could be tracked – spillover at an ice-covered reservoir in winter with low to no snow pack. The alga that produced the odor grew just under the ice layer of the reservoir. At this time of year, when conditions were right, the treatment plant made sure it had potassium permanganate ready for treatment.

Some insight into algae growth dynamics can be obtained by measuring water temperature, pH, and dissolved oxygen. Photosynthesis of algae affects the pH and dissolved oxygen of water in diurnal cycles (as shown in Figure 5-4).

Measurements can be made using a data sonde, which is a continuously operating, remotely stationed and submerged metering device. A data sonde can be submerged in water for as long as two weeks to one month to collect readings on pH and dissolved oxygen and other parameters. Gathering such historical data from various sampling stations obtains a better understanding of algal activity, which can then be better correlated to the occurrence of T&O problems.

Chlorophyll a (chl-a) and other pigments are often measured as estimates of the abundance of the total algal community. These analyses, which can provide a relatively simple, useful gauge of abundance, present limitations. There is often a poor correlation between chl-a and total algal biomass, mainly because the cell content of chl-a and other pigments can change under different light and nutrient regimes and with different taxonomic groups and cell sizes (e.g., large vacuolar cells vs. small nanoplankton).

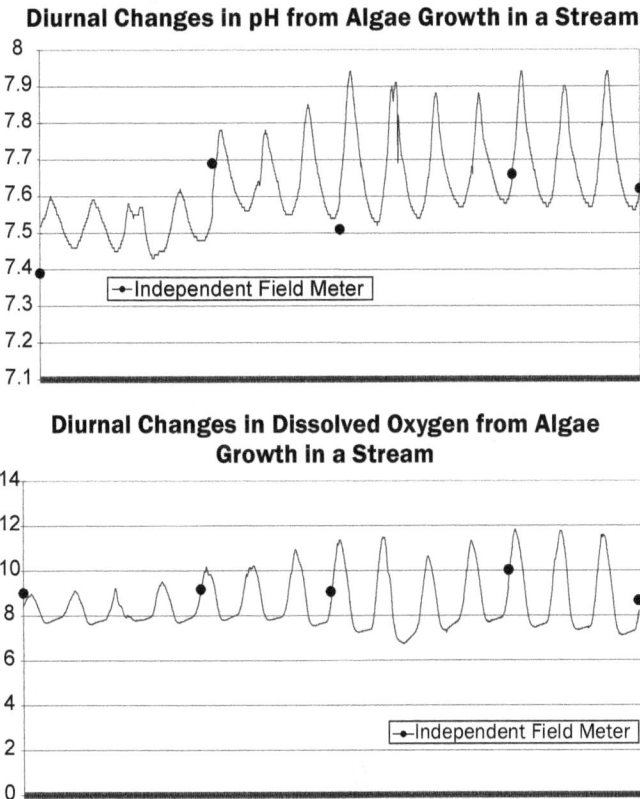

Courtesy of Gary Burlingame.

Figure 5-4 Example of diurnal changes in pH and dissolved oxygen as caused by algal activity in a source water supply

New high-tech methods have been developed in an attempt to automate this process and/or to collect real time, inline, or more spatially extensive and frequent records. These typically involve image processing (flow cytometry) or fluorescence signals to estimate the relative abundance of different taxonomic groups. These methods require careful calibration. They can provide a useful screening but have limitations in what they measure and to what extent the results can be correlated with algal species abundance.

Bibliography

American Public Health Association. 1905. *Standard Methods of Water Analysis*, 1st ed. American Public Health Association, Washington, D.C.

American Public Health Association, American Water Works Association, and Water Environment Federation. 2005. *Standard Methods for the Examination of Water and Wastewater*. 21st ed. American Public Health Association, Washington, D.C.

Burlingame, G.A., Dann, R.M., and Brock, G.L. 1986. A Case Study of Geosmin in Philadelphia's Water. *JAWWA* 78:3:56.

Burlingame, G.A., Muldowney, J.J., and Maddrey, R.E. 1992. Cucumber Flavor in Philadelphia's Drinking Water. *JAWWA* 84:8:92

Burlingame, G.A., Dietrich, A.M. and Whelton, A. J. 2007. Understanding the Basics of Tap Water Taste. *JAWWA* 99:5:100.

Desrochers, R. 2008. Sensory analysis in the water industry. *JAWWA* 100:10:50.

Dietrich, A.M. 2006. Aesthetic Issues for Drinking Water. *Journal Water and Health*, 4:1.

Dietrich, A.M., Hoehn, R.C., Burlingame, G.A., and Gittelman, T. 2004. *Practical Taste-and-Odor Methods for Routine Operations: Decision Tree*. Water Research Foundation, Denver, Colo.

Dietrich, A.M., Whelton, A., Hoehn, R., Anderson, R., and Wille, M. 2004. Attribute rating test for sensory analysis. *Water Science and Technology*, 49:9:61.

Dietrich, A.M., Burlingame, G.A., Vest, C., and Hopkins, P. 2004. Rating method for evaluating distribution-system odors in comparison to a control. *Water Science and Technology*, 49:9:55.

Graham, M., Najm, I., Simpson, M., MacLeod, B., Summers, S., and Cummings, L. 2000. *Optimization of Powdered Activated Carbon Application for Geosmin and MIB Removal*. Water Research Foundation, Denver, Colo.

Khiari, D., Barrett, S., Chinn, R., Bruchet, A., Piriou, P., Matia, L., Ventura, F., Suffet, I., Gittelman, T., and Luitweiler, P. 2002. *Distribution Generated Taste and Odor Phenomena*. Water Research Foundation, Denver, Colo.

Krasner, S.W., McGuire, M.J., and Ferguson, V.B. 1985. Tastes and odors: the flavor profile method. *JAWWA* 77:34.

Lawless, H.T., and H. Heyman. 1998. *Sensory Evaluation of Food: Principles and Practice*. Chapman and Hall, New York, N.Y.

Lund, J.W.G., Kipling, C. and Le Cren, E.D. 1958. The inverted microscope method of estimating algal numbers and the statistical basis of estimations by counting. *Hydrobiol.*, 11:143.

Mackey, E.D., Baribeau, H., Fonesca, A.C., Davis, J., Brown, J., Boulos, L., Crozes, G.F., Piriou, P., Rodrigues, J.M., Fouret, M., Bruchet, A., and Hiltebrand, D.J. 2004. *Public Perception of Tap Water Chlorinous Flavor.* Water Research Foundation, Denver, Colo.

Mackey, E.D., Suffet, I.H., and Booth, S.D.J. 2010. *A Decision Tool for Earthy/Musty Taste and Odor Control.* Water Research Foundation, Denver, Colo.

Mallevialle, J. and Suffet, I.H. (eds.) 1987. *Identification and Treatment of Tastes and Odors in Drinking Water.* Water Research Foundation, Denver, Colo.

McGuire, M., Graziano, N., Sullivan, L. Hund, R., and Burlingame, G. 2004. *Water Utility Self-Assessment for the Management of Aesthetic Issues.* Water Research Foundation, Denver, Colo.

Meilgaard, M., G.V. Civille, and B.T. Carr. 2007. *Sensory Evaluation Techniques*, 4th ed. CRC Press, Boca Raton, Fla.

Noblet, J., Schweitzer, L., Ibrahim, E., Stolzenbach, K.D., Zhou, L., and Suffet, I.H. 1999. Evaluation of a taste and odor incident on the Ohio River. *Water Science and Technology*, 40:6:185.

Ömür-Özbek; P. and Dietrich, A.M. 2008. Developing hexanal as an odor reference standard for sensory analysis of drinking water. *Water Research*, 42:2598.

Paxinos, R. and Mitchell, J.G. 2000. A rapid Utermöhl method for estimating algal numbers. *Journ. of Plankton Research*, 22:12:2255.

Rashash, D.M.C. and Gallagher, D.L. 1995. An evolution of algal enumeration. *JAWWA*, 87:127.

Rigal, S. 1992. The use of organoleptic investigations to evaluate the quality of materials in contact with drinking water. *Water Science and Technology*, 25:41.

Romero, J., Ventura, F., Caixach, J., Rivera, J., Godé, L.X., and Ninerola, J.M. 1998. Identification of 1,3-dioxanes and 1,3-dioxolanes as malodorous compounds at trace levels in river water, groundwater and tap water. *Environmental Science and Technology*, 32:206.

Sandgren, G.D. and Robinson, J.V. 1984. A stratified sampling approach to compensating for nonrandom sedimentation of phytoplankton cells in inverted microscope settling chamber. *Br. Phycol. Journ.*, 19:67.

Suffet, I., Mallevialle, J., and Kawczynski, E. (eds.). 1995. *Advances in Taste and Odor Treatment and Control*. Water Research Foundation, Denver, Colo. and Lyonnaise des Eaux, Paris, France.

Taylor, W.D., Losee, R.F., Torobin, M., Izaguirre, G., Sass, D., Khiaria, D., and Atasi, K. 2006. *Early Warning and Management of Surface Water Taste-and-Odor Events*. Water Research Foundation, Denver, Colo.

Whelton, A.J., Dietrich, A.M., Gallagher, D.L., and Roberson, J.A. 2007. Using customer feedback for improved water quality and infrastructure monitoring. *JAWWA*, 99:11:62.

Whelton, A.J., Dietrich, A.M., Burlingame, G.A., and Cooney, M.F. 2004. *Detecting Contaminated Drinking Water: Harnessing Consumer Complaints*. In Proc. AWWA Water Quality Technology Conference, San Antonio, Texas.

Willen, E. 1976. A simplified method of phytoplankton counting. *Br. Phycol. Journ.*, 11:265.

Appendix A

From *Journal AWWA*, May 2007

Understanding the basics of tap water taste

by Gary A. Burlingame, Andrea M. Dietrich, Andrew J. Welton

In many ways tap water is similar to food and beverage products sold at the grocery store. It has a shelf life (i.e., changes in quality can occur over time), a preservative (i.e., a disinfectant residual), and packaging (e.g., pipes, plumbing, storage tanks). As long as we supply drinking water through a public water supply system, we have the responsibility to consider its acceptability for human consumption—even though only a small percentage of tap water is used for drinking. Although a lack of acceptance of tap water flavor will not likely affect water production, it can be a major factor in public trust and confidence. Therefore, having an understanding of how the sensory properties of tap water are perceived is important for customer relations (Dietrich, 2006).

Despite the strong relationship between taste and acceptability, the connection between drinking water taste, water quality, and water treatment has received little attention. As water treatment techniques become more effective at removing constituents from water— such as with desalination and membrane processes—the effect of water treatment on water taste is becoming more important.

Anions and cations, including the hydrogen ion, are largely responsible for the taste sensations at the cellular level of the taste buds. In water, anions and cations occur as a multifaceted "soup" of free ions, complexes, and even particulate matter. Ions are present simultaneously and interact depending on concentration, pH, pE, temperature, and thermodynamic stability constants. For example, copper can be present as the free cupric ion Cu^{+2}; it can bind with anions to form hydroxo complexes (e.g., $CuOH^+$), carbonato complexes (e.g., $CuCO_3{}^0$), and chloride complexes (e.g., $CuCl^+$); or it can occur as the precipitate copper II hydroxide. The taste perception of these different ion species has yet to be thoroughly investigated or understood. Thus, references on what affects taste vary between reporting levels of ions and levels of minerals and salts.

A literature review was conducted to explore how "taste" is determined, how the levels of minerals (cations and anions) that occur in natural waters and saliva influence the taste of water, how existing standards address taste, how water treatment can affect taste, and whether more research is needed. The results of this review will provide drinking water practitioners, treatment designers, researchers, and regulators with a better understanding of how their decisions can affect the taste of tap water. Included is an overview of the human senses with regard to water flavor.

Terminology is key to understanding taste

It is necessary to understand how people perceive the taste of water in order to optimize water taste, and, for scientists to communicate their findings, terminology must be clarified. The "instruments" of detection, in this case, are found in the human olfactory and gustatory systems. The terminology associated with the human senses has been described by other scientists in fields outside the drinking water industry, and key terms are explained here in detail.

Flavor. This term describes the sum of human sensations created by food, beverage, and even tap water (Lawless & Heymann, 1998). It is caused by three separate sensations, namely taste, odor, and feelings. Human response to taste or odor sensations can be characterized in two ways: with a description by terms, such as "sweet" or "musty," and an intensity rating such as "weak" or "strong." Any product that is tasted or smelled can be sensed as a combination of several descriptive qualities, each having its own intensity.

Odor. The sense of smell (olfactory ssensitivity or olfaction) provides the human response to odors (Köster, 2002; Lawless & Heymann, 1998) and deserves discussion because odor is often associated with taste. Odors are perceived when substances enter the nasal cavity directly through the nostrils and when they are released during mouth movements created during tasting and swallowing (retronasal). Odor qualities are innumerable and are typically characterized by past associations (e.g., rose perfume, onion bagel, rubber tires, and ammonia cleaner). Odors evoke memory—thus experiences and memory are important in an individual's olfactory capability. Unfortunately, the ability to smell decreases with age much more so than does taste sensitivity, and anosmia (the inability of some people to detect all or some odors) can occur for short and long periods of time because of various olfactory disorders (Doty et al., 1984).

Taste. Taste is a very general term that is used in a variety of ways, such as to denote general tasting or the whole of the sensory experience when a food or beverage is consumed. Technically, taste has a very specific meaning (Gilbertson et al., 2006; Matsuo, 2000; Smith & Margolskee, 2001; Lawless & Heymann, 1998), and is referred to as "gustatory sensitivity." Four basic taste sensations (tastants) are widely recognized: sour, salty, bitter, and sweet. A fifth taste called umami has been proposed, but it is not widely recognized (Smith & Margolskee, 2001; Yamaguchi, 1991). Umami—Japanese for "meaty"—is the taste sensation caused by amino acids, proteins, and monosodium L-glutamate.

A chemist understands how a pH electrode detects hydrogen ion activity, and a biologist understands how a microscope brings forth the hidden world of microorganisms. In the case of "taste," the sensor to better understand is the taste bud. Taste buds containing taste receptors are located in adults primarily on the tongue but are also found throughout the mouth and on the back of the throat (pharynx). A taste bud consists of a group of cells with microvilli that extend into a pore that is open to the mouth. When chemicals contact these cells, impulses are sent to the brain by one of three nerve pathways. One pathway passes by the ear, which is why ear problems can some-times cause alterations in taste (Lawless & Heymann, 1998). Some people cannot taste anything, a condition called aguesia. People can be grouped according to the frequency of taste buds in their mouth (Lawless & Heymann, 1998): "super-tasters" (hypersensitive people) have the greatest frequency, followed by "tasters" (sensitive people), and "nontasters" (insensitive people).

Historically, and still reproduced in some recent publications, a map of taste perceptions on the tongue depicts regions of specific sensitivities to tastes such as sour on the sides of the tongue, or bitter on the back and top surface of the tongue. Although there are regions of greater taste sensitivity, each of the basic tastes can be perceived across the entire surface of the tongue and throughout many regions of the mouth (Smith & Margolskee, 2001; Lawless & Heymann, 1998).

From a physical or biochemical perspective, the perception of sour may be the simplest of taste sensations—the stimulating components are hydrogen ions, which pass through the membrane of the taste cell to initiate the sensation of sourness (Smith & Margolskee, 2001). This sensation can reduce a person's sensitivity to other compounds (which is also true for other tastes), though the interference is usually short-lived. Sour is not a sensation commonly used to describe drinking water.

The salty taste is produced primarily by the sodium ion (Smith & Margolskee, 2001; Yamaguchi, 1991). Thus when we salt our food with sodium chloride, it is the sodium and not the chloride ion that stimulates taste. Like the hydrogen ion of sourness, the sodium ion passes through the cell membrane and initiates the perception of salty. Many other inorganic salts evoke either bitter or sweet tastes, not salty. A salty water taste is most commonly associated with the taste of seawater or brackish water but can also be associated with waters of high total dissolved solids (TDS).

High-molecular-weight inorganic salts usually taste bitter (Smith & Margolskee, 2001). Bitter can be caused by a variety of compounds—organic (such as caffeine) and inorganic (such as magnesium chloride and calcium chloride)—and can linger for a long time. Some people are overly sensitive to bitter sensations, while others are insensitive (Fast et al., 2002; Lawless & Heymann, 1998). Water with a high level of some metals, such as copper, can have a bitter taste (Cuppett et al., 2006).

In our everyday experiences, the sweet sensation is typically caused by sugars (Smith & Margolskee, 2001). Some inorganic salts, notably lead salts, taste sweet. Research has shown that at the perceptual level sugars can interfere with other taste sensations (Lawless & Heymann, 1998); thus sugar inhibits to some degree the bitterness of coffee or sourness of lemonade.

Aftertaste is a quality that has played an important role in customer acceptance of foods and beverages. It is defined as the sensation that lingers in one's mouth after the food or beverage is swallowed, and it can last for more than 30 seconds.

Finally, the serving temperature of water can have a strong influence on drinking water acceptability and taste. The temperature for optimum taste perception appears to be in the range of 20–40°C, in which thermal sensitivities such as "hot" and "cold" are minimal. Although tastes may be optimized at a temperature near normal body temperature, people prefer to drink chilled water (Brunstrom et al., 1997).

Table A-1 Major anions and cations of artificial saliva

	Concentration—mg/L	
Constituent	Schafer & Kandbiglari (2003)	Preetha & Banerjee (2005)
Sodium—Na^+	396	396
Potassium—K^+	704	629
Calcium—Ca^{+2}	55	47
Chloride—Cl^-	1,178	1,151
Inorganic total dissolved solids	2,700	2,200

Mouthfeel. Terminology is key in communicating about the human sense of taste. Confusion can readily appear, such as in describing "flavor," which involves odor, taste, and feeling factors. Feelings (physical sensations) or "mouthfeel" evoked by some chemicals in the nose or mouth result from stimulation of the trigeminal nerves. Trigeminal nerve endings are scattered throughout the nose, mouth, pharynx, larynx, and on the tongue (Smith & Margolskee, 2001; Lawless & Heymann, 1998). Such feelings include pain, thermal (hot and cold), pressure, and chemical (astringent, drying, slick) from irritating substances like ammonia, horseradish, onion, chili pepper, menthol, carbonation, and high concentrations of salts. Feelings can allow totally anosmic people to sense some odorants like ammonia and spearmint. Sensations produced by irritants in the throat can elicit coughing, whereas sensations in the nose can elicit sneezing. Overall, compound concentrations required for eliciting feelings are greater than those required for eliciting taste or odor perception (Lawless & Heymann, 1998). In addition, for some irritants, the chemical sensations are likely to persist for a time after exposure, e.g., the burn from hot peppers. In drinking water, metals such as zinc, copper, iron, and manganese can cause astringency. Some metals can also cause an electric sensation (Lawless et al., 2005; 2003).

Effect of saliva. It might seem logical to think that a glass of water is optimal for taste when everything else in the water besides H_2O—all the "contaminants"—have been removed. However, a better understanding of taste shows us that good-tasting water has "ingredients," such as minerals, that make it taste good. These ingredients form an optimal recipe according to each person's saliva content. The taste of water is largely affected by the background conditions of the human taste buds, which are constantly bathed in saliva (Lawless & Heymann, 1998). According to Matsuo (2000), "Water is the most important salivary constituent for the maintenance of taste acuity." Saliva pH is typically 6.5–7.4 (Pedersen et al., 2002) and contains electrolytes important to the body, such as cations (sodium, magnesium, potassium, and calcium) and anions (chloride, bicarbonate, phosphate; Pedersen et al., 2002; Smith & Margolske, 2001; Matsuo, 2000). Water makes up about 99% of saliva, and proteins and salts make up the rest (Pedersen et al., 2002). Trace (micromolar) amounts of various other chemicals can also be found in saliva (Tenovou, 1989). Thus, individual differences in saliva content and volume, such as would be affected by diet, affect a person's sensitivity to taste.

Our taste sensations are triggered in reference to what is already in our mouths. Saliva solubilizes substances we ingest and influences what reaches the taste buds. Salivary proteins and peptides can bind cations and flavor components (Friel & Taylor, 2001; Argarwal & Henkin,1987). Because saliva contains bicarbonates, phosphate, and proteins, it acts as a buffer. Tastes such as bitter and sour, as well as a low pH, induce increased saliva flow (Pedersen et al., 2002; Matsuo, 2000; Guianard et al., 1998; Christensen et al., 1987). As saliva flow increases, the concentrations of proteins, sodium, calcium, chloride, and bicarbonate also increase (Pedersen et al., 2002). A change in saliva flow can change sensitivity to taste (Matsuo, 2000). For example, sodium in saliva increases in concentration when saliva production increases, which in turn raises the level that sodium chloride must reach for a person to detect its presence in the oral cavity (Delwiche & O'Mahony, 1996).

Adults can produce 500 mL or more of saliva in a day. One important reason for production of saliva is to replace minerals lost from tooth enamel during wear. Saliva needs to be supersaturated with the ions important to teeth (calcium, phosphate, hydroxide ions). Therefore, the rate of saliva flow affects the levels of calcium, phosphate, and bicarbonate in the mouth. Information on the importance of saliva can be found at various sources.[1-3]

Studies have reported on recipes for saliva, which included the following substances that could occur in water (Preetha & Banerjee, 2005; Schafer & Kandbiglari, 2003; van Ruth & Roozen, 2000): potassium chloride, sodium chloride, magnesium fluoride, calcium chloride, photassium phosphate, and magnesium chloride. These artificial saliva recipes are compared in Table A-1. Although no standardized recipe for saliva exists, it is clear that typical recipes have high TDS values (>2,000 mg/L), potassium (>500 mg/L), sodium (>300 mg/L), and chloride (>1,000 mg/L), with either carbonates or phosphates as the major anions to provide buffer capacity.

The salty taste of sodium chloride occurs when its levels in the ingested substance exceed those in the saliva surrounding the taste buds (Pedersen et al., 2002; Matsuo, 2000). Seawater contains relatively high levels of minerals (Conry, 2004) and so, in comparison to saliva, tastes strong to humans. Although potassium does not appear to affect taste directly, at the levels found in water, for example, it could play an indirect role by enabling the taste buds to become more active. This is because potassium is a key ion in the communication between nerve cells (Matsuo, 2000).

Table A-2 Cations and anions in water that can affect taste, and their typical levels in natural freshwaters

Anion or Cation	Common Forms	Levels in Freshwater—mg/L
Calcium—Ca^{+2}	$Ca(HCO_3)_2$, $CaSO_4$, $CaCl_2$, Ca_3SiO_5, Ca_2SiO_4, $CaCO_3$	As much as 600
Magnesium—Mg^{+2}	$MgCO_3$, $Mg(HCO_3)_2$, $MgSO_4$, $MgCl_2$, $CaMg(CO_3)_2$	As much as several hundred
Sodium—Na^+	$NaHCO_3$, Na_2SO_4, Na_2CO_3, $NaCl$	As much as 1,000
Silica—Si^{+2} or SiO_2	H_4SiO_4, Ca_3SiO_5, Ca_2SiO_4, $Mg_2(H_2O)_2(SiO_2)_3$	Ranges from 1 to 30, although levels as high as 100 can be common
Potassium—K^+	$KHCO_3$, K_2SO_4, KCl, K_2HPO_4, K_2CO_3	Generally <10
Bicarbonate—HCO_3^-	$NaHCO_3$, $Mg(HCO_3)_2$, $Ca(HCO_3)_2$, $KHCO_3$	Commonly <1,000
Chloride—Cl^-	$NaCl$, $CaCl_2$, KCl, $MgCl_2$	Commonly <10
Sulfate—SO_4^{-2}	$CaSO_4$, Na_2SO_4, $MgSO_4$, $Al_2(SO_4)_3$, KSO_4	Commonly <100
Carbonate—CO_3^{-2}	$CaCO_3$, $MgCO_3$, K_2CO_3, Na_2CO_3, $CaMg(CO_3)_2$	Commonly 0 in surface waters or <10 in groundwaters
Total dissolved solids	Combination of all of the above	<3,000 (freshwater should have <1,000)

Source: van der Leeden et al., 1990

Natural minerals contribute to taste

Minerals found naturally in water (Table A-2; van der Leeden et al., 1990) can play a major role in its taste. Cations such as calcium and sodium can affect drinking water taste and minerals such as calcium bicarbonate, calcium sulfate, and calcium chloride are likely to be found as dissolved constituents in tap water. How important are minerals to taste? Just read the label on some bottles of purified drinking water. The label on one purified drinking water states: "minerals added for taste." It further explains: "Contains purified water and specifically selected minerals in nutritionally insignificant amounts for clean, fresh taste (sodium bicarbonate, magnesium chloride, calcium chloride, and sodium sulfate)." Although a person might notice a difference in taste with waters of different mineral contents, it may not be until the actual minerals significantly exceed a person's taste thresholds that the person actually recognizes the reason for the difference in taste (USEPA, 1979).

Cations. Calcium is very common in water. The taste threshold concentration (TTC) for calcium, depending on its associated anion, has been found to be 100–300 mg/L (WHO, 2004b). The TTC is the concentration at which a person can detect a taste, which can vary for many reasons, including cultural differences or experiences and genetics. TTCs are highly dependent on the ions, pH, and temperature of the water (WHO, 2004b). Most drinking waters have calcium levels less than 50 mg/L (van der Leeden et al., 1990). Soft waters tend to have no effect on taste. Good-tasting water has been reported to have a total hardness between 10 and 100 mg/L. This is due largely to calcium hardness, although it's still unclear whether calcium contributes directly to drinking water taste (Koseki et al., 2003). Soft waters have a hardness (as calcium carbonate) between 0 and 60 mg/L; moderately hard waters between 61 and 120 mg/L; and hard waters between 121 and 180 mg/L (van der Leeden et al., 1990). Mineral waters can have levels of calcium above 150 mg/L (van der Aa, 2003). Calcium in seawater can be found at about 400 mg/L (Desai, 2004).

The level of magnesium that people report as having an objectionable taste is affected by the anion with which it is associated. Magnesium is typically associated with anions such as carbonate, bicarbonate, sulfate, and chloride. Magnesium can be detected by consumers at 100–500 mg/L and can impart an astringent or bitter taste (Lockhart et al., 1955). However, water containing magnesium salts at 1,000 mg/L has been considered acceptable (Bruvold & Pangborn, 1966). Magnesium in tap water can reach up to 120 mg/L, although most tap waters fall below 20 mg/L (van der Leeden et al., 1990). In comparison, some mineral waters have levels above 50 mg/L (van der Aa, 2003). Magnesium can be found in seawater at about 1,000 mg/L (Desai, 2004).

Sodium is commonly found in water from natural sources as well as from its widespread use in society. Most drinking waters contain <50 mg/L (van der Leeden et al., 1990) and "mineral" waters are >200 mg/L (van der Aa, 2003). The TTC for sodium varies from 30 to 460 mg/L depending on many factors, such as serving temperature and the age and diet of the person being exposed (USEPA, 2003a). The World Health Organization (WHO) recommends that sodium not exceed 200 mg/L to avoid a salty taste (WHO, 2004b). The US Environmental Protection Agency (USEPA) recommends that sodium not exceed 30–60 mg/L to avoid any taste effects and to help in reducing the sodium intake for those on a low-sodium diet and concerned about hypertension (USEPA, 2003a). Wiesenthal et al (2007) found the taste threshold for sodium chloride (in deionized water at pH 8) to be about 650 mg/L. Although human saliva contains sodium chloride and sodium bicarbonate (Matsuo, 2000), sodium is responsible for the salty taste (Pedersen et al., 2002; Smith & Margolskee, 2001; Matsuo, 2000). Sodium in seawater can be found at about 10,000 mg/L (Desai, 2004).

Potassium is important at the cellular level of the taste buds (Matsuo, 2000). Potassium is in human saliva associated with chloride, bicarbonate, and the phosphate anion (Matsuo, 2000). Potassium chloride acts similar to sodium chloride in taste effects. Most tap water has <5 mg/L of potassium (van der Leeden et al., 1990). Potassium can be found in seawater at about 200–400 mg/L (Desai, 2004).

Anions. Anions such as bicarbonate, chloride, and sulfate can affect taste. At neutral pH, bicarbonate is more important than carbonate and helps keep cations in solution. Carbonate concentrations become more prominent at pH values >8.3. Most tap waters contain <1 mg/L carbonate (van der Leeden et al., 1990). Together, bicarbonate and carbonate are the major ions contributing to alkalinity and buffer capacity in water (Stumm & Morgan, 1970). The taste of bicarbonates is preferred over carbonates, but a pH change is necessary when shifting between these two species.

Bicarbonate is the major species at pH values of 6.3–8.3 and is associated with the cations sodium, calcium, magnesium, and potassium. Mineral waters have bicarbonate levels >600 mg/L (van der Aa, 2003), whereas most tap waters contain <150 mg/L of bicarbonate (van der Leeden et al., 1990).

The TTC for chloride is 200–300 mg/L (WHO, 2004b; Lockhart et al., 1955; Ricter & MacLean, 1939). Potassium chloride and magnesium chloride are less noticeable in taste than sodium chloride and calcium chloride (taste thresholds are around 250–525 mg/L; USEPA, 1979). Most tap waters contain <50 mg/L Cl⁻ (van der Leeden et al 1990). Increased chloride levels in water in the presence of sodium, calcium, potassium, and magnesium, can cause objectionable water taste. Mineral waters have chloride levels above 200 mg/L (van der Aa, 2003). Chloride can be found in seawater at about 19,000 mg/L (Desai, 2004).

Sulfate has minimal impact on taste at low levels, but the TTC is about 250 mg/L for sodium sulfate and about 1,000 mg/L for calcium sulfate (WHO, 2004b). Magnesium sulfate is less noticeable in taste than sodium sulfate or calcium sulfate (USEPA,1979). Sulfate, though, has a laxative effect when it exceeds 500–1,000 mg/L, depending on the presence of other ions, and can impart a salty-type taste when it exceeds 250 mg/L (USEPA, 2003b; 1979). Common aqueous cations associated with sulfate are calcium, sodium, magnesium, aluminum, and potassium (e.g., in such forms as calcium sulfate, sodium sulfate, magnesium sulfate, aluminum sulfate, and potassium sulfate, respectively). Tap water should contain low levels of sulfate. Sulfate in most tap waters is <100 mg/L (van der Leeden et al., 1990) and in mineral waters would be >200 mg/L (van der Aa, 2003). Sulfate in seawater can be found at about 2,000–3,000 mg/L (Desai, 2004; USEPA, 2003b).

Silica, associated with calcium and magnesium (Stumm & Morgan, 1970), has an unclear relation to taste of tap water. Most tap waters have <30 mg/L silicon dioxide (van der Leeden et al., 1990).

Temperature and pH. Water temperature and pH are also important factors that affect the taste of water (Zellner et al., 1988; Pangborn & Bertolero, 1972). Tap water's serving temperature can range from 4 to about 40°C. Brunstrom et al (1997) reported that cold water (refrigeration temperature) induces more saliva flow, which in turn provides relief of dry mouth and other sensations associated with thirst. Thus, cold water is more refreshing and pleasant.

The pH of drinking water also affects taste and should fall between 6.5 and 8.5 (USEPA, 1979). This also encompasses the typical pH range of saliva (Pedersen et al., 2002). The taste effect of pH may be due to changes in ion speciation rather than changes in hydrogen ion concentration. One study demonstrated that varying the pH of distilled water from a pH of 7 to a pH of 9 did not affect human perception (Cuppett et al., 2006). Water pH influences bicarbonate and carbonate levels as well as the formation of various metal ion complexes.

Drinking water can provide essential minerals

A desirable tap water has no smell, no lingering aftertaste, and no undesirable mouthfeel sensations (such as drying or tongue coating). It does not taste unacceptably sweet, sour, bitter, or salty, nor does it taste flat. Most people agree that a good-tasting tap water has a balance of minerals, chilled water temperature, and near-neutral pH.

It gives a refreshing sensation. Water must taste acceptable because its intake is essential to human health. With respect to the minerals in water that balance its taste, the most important ones for human health are sodium, chloride, potassium, and sulfate (NAS, 2005).

WHO has identified calcium, sodium, chloride, magnesium, iron, zinc, copper, chromium, cobalt, molybdenum, and selenium as essential elements for human health (WHO, 2004a). Boron, manganese, nickel, silica, phosphorus, potassium, and vanadium are also quite important from a human health perspective. These minerals help our bodies maintain strong bones, send electrical impulses, build muscle, lower blood pressure, and digest food. People vary in their consumption of water, which can be influenced by the water's quality and acceptability for drinking, which in turn affects how beneficial the water is for supplying essential elements (Szlyk et al., 1990, 1989; Hubbard et al., 1990). Where individuals obtain their minerals greatly depends on diet, water source, and other factors. Drinking water can provide a significant source of calcium and magnesium.

How ingredients contribute to aesthetic standards

Safe drinking water standards exist to prevent acute and chronic health risks. The USEPA also sets standards for aesthetics—secondary maximum contaminant levels (SMCLs)—as shown in Table A-3. The difficulty in setting national standards for taste and odor compounds, based on public perception and the methods available to evaluate public sensitivity, was addressed by the AWWA Taste and Odor Committee (2002).

Metals. Several metals have SMCLs (Table A-2): copper should not exceed 1 mg/L, iron should not exceed 0.3 mg/L, and zinc should not exceed 5 mg/L. These three are the most common metals found in "metallic" tasting water. Although a common condition is rusty water from high levels of iron, metals that leach from plumbing into water can be tasted at concentrations that do not impart an off-color to the water. Lim & Lawless (2005) stated, "Although the perception of metallic compounds is not well understood, the term 'metallic taste' is commonly used to describe many sensations evoked by metallic compounds." A "metallic" taste includes tactile sensations and retronasal cues.

Metallic and astringent tastes, noticed as a lingering aftertaste, more often arise from the corrosion or leaching of piping materials, such as copper and iron. Corrosion of plumbing materials can be enhanced by stagnant water conditions (such as when a house is closed up for long periods of time) or by faulty plumbing. Chemical injection system failures and backflow through cross connections can also introduce high levels of metals (such as from a wet fire sprinkler system) or chemicals that corrode pipe. The backflow of carbon dioxide from a soda fountain system and its resultant leaching of copper from the potable water supply piping can cause a metallic taste.

Iron can be present in soluble, suspended, or hydroxide forms. Most drinking water standards for iron are 0.3 mg/L to avoid laundry staining, turbidity, and color formation (WHO, 2004b), but iron can also impart a "metallic" taste (Table A-2). Typical levels in natural freshwaters are generally less than 0.5 mg/L although some groundwaters can have up to 10 mg/L (van der Leeden et al., 1990).

Copper, zinc, and manganese can affect the taste of water (WHO, 2004b). When the copper concentration exceeds about 4 mg/L, gastrointestinal upset, a bitter taste, and toxicity can occur (Dietrich et al., 2004b; Olivares et al., 2001; Pizzaro et al., 1999; Cohen et al., 1960). Although USEPA's SMCL for copper is 1 mg/L, a majority of people can detect its presence in water at concentrations below this—even as low as 0.1 mg/L (Cuppett et al., 2006). Zinc can be detected as zinc sulfate at 4 mg/L Zn (WHO, 2004b). Zinc salts have been found to have a lingering astringency (Keast et al., 2003). Manganese has been found to impart an astringent taste and has a TTC of about 0.05 mg/L (WHO, 2004b). Manganese levels should be low in tap water. Natural freshwaters generally have manganese concentrations of 0.20 mg/L or less, although some groundwaters can have up to 10 mg/L (van der Leeden et al., 1990).

TDS. The SMCL for TDS is 500 mg/L. TDS is a measure of the total ion concentrations, including cations such as calcium, magnesium, potassium, sodium, aluminum, iron, and manganese and anions such as bicarbonate, carbonate, chloride, sulfate, and nitrate. Reverse osmosis– (RO–) treated water and distilled water can have TDS values <10 mg/L.

TDS varies significantly depending on the water source and local geology (van der Aa, 2003). Typical tap water ranges for low, moderate, and high TDS waters are <100, 101–250, and 251–500 mg/L concentrations, respectively. Water close to zero TDS has a flat taste. US drinking water palatability assessments have found the following TDS ratings (Bruvold & Daniels, 1990): 80 mg/L—excellent, 81–450 mg/L—good, 451–800 mg/L—fair, 801–1,000 mg/L—poor, and >1,000 mg/L—unacceptable to taste. Most tap waters have TDS levels below 500 mg/L (van der Leeden et al 1990) at which point TDS has no undesirable effects on taste.

Table A-3 USEPA SMCL for chemicals that can affect the taste of drinking water

Ingredient	SMCL—mg/L	Effects on Taste
Chloride	250	Salty (TTL: 200–300 mg/L)
Copper	1.0	Metallic
Iron	0.3	Metallic
Manganese	0.05	Bitter, metallic
pH	6.5–8.5 desired range	Low pH—bitter, metallic High pH—slippery feel, soda taste
Sulfate	250	Salty (TTL: 250–1,000 mg/L)
Total dissolved solids	500	Salty
Zinc	5	Metallic

Source: USEPA, 1979
SMCL—secondary maximum contaminant levels, TTL—taste threshold level, USEPA—US Environmental Protection Agency

Table A-4 Levels of minerals in mineral waters

Minerals	Levels Can Exceed—mg/L
Calcium—Ca^{+2}	150
Magnesium—Mg^{+2}	50
Sodium—Na^+	200
Chloride—Cl^-	200
Sulfate—SO_4^{-2}	200
Total hardness—as $CaCO_3$	200
Total dissolved solids	500

Source: van der Aa, 2003

Mineral waters. Mineral waters clearly have a different taste than typical tap waters. The US Food and Drug Administration's standard for the identity and labeling of bottled waters defines "mineral water" as containing at least 250 mg/L TDS. The mineral content of waters is usually defined by the levels of sodium, calcium, sulfate, and magnesium, particularly by hardness, TDS, and chloride (van der Aa, 2003). Table A-4 summarizes the levels of minerals that can be found in mineral waters (van der Aa, 2003).

Table A-5 Minerals in common rock

Rock Name	Mineral Formulas
Feldspar	$KAlSiO_3$, $NaAlSi_3O_3$, $CaAl_2Si_2O_3$
Quartz	SiO_2, $CaMg(SiO_3)_2$
Pyroxene	$(Mg,Fe)SiO_3$, $(Al,Fe)SiO_3$
Hornblende	$Ca_2Mg_5Si_8O_{22}(OH)_2$, $Ca(Mg,Fe)_5Si_8O_{22}(OH)_2$
Mica	$KAl_2(Si_3Al)O_{10}(OH)_2$, $K_2(Mg,Fe)_6(SiAl)_8O_{20}(OH)_2$
Olivine	$(Mg,Fe)_2SiO_4$
Carbonate	$CaMg(CO_3)_2$, $FeCO_3$
Kaolinite	$Al_2Si_2O_5(OH)_4$
Ferric oxide	Fe_2O_3, $2Fe_2O_3 \cdot 3H_2O$

Source: Chien & Wan, 1999

Table A-6 Recipe for hard and soft natural waters* and median levels of minerals in 100 large cities in the United States†

Ions	Hard Water Level mg/L	Soft Water Level mg/L	Median Level mg/L
Calcium—Ca^{+2}	49.0	10.6	26
Magnesium—Mg^{+2}	9.9	1.5	6.2
Sodium—Na^+	15.3	5.8	12
Potassium—K^+	4.1	1.0	1.6
Bicarbonate—HCO_3^-	122.6	23.4	46
Chloride—Cl^-	24.3	9.9	13
Sulfate—SO_4^{-2}	116.2	22.0	26
Nitrate—NO_3^-	6.2	1.8	0.7
Phosphate—H_2PO4^-	2.9	0	Not reported
Hydronium—H^+	3.24×10^{-6} at pH 8.49	1.48×10^{-5} at pH 7.83	3.16×10^{-5} at pH 7.50

*Source: Smith et al., 2002
†Source: van der Leeden et al., 1990

Where the minerals come from

Minerals or inorganic chemicals enter natural waters by the weathering, erosion, or disturbance of rock and soil. Table A-2 summarizes the levels of minerals in natural freshwaters. Table A-5 shows the principle minerals that make up common rocks, which weather and contribute to the minerals in natural freshwaters.

The alkalinity, or carbonate/bicarbonate chemistry of water, is important in determining the availability and occurrence of minerals in water (Stumm & Morgan, 1970). Alkalinity is the acid-neutralizing capacity of water and is based on carbonate chemistry. At a neutral pH, bicarbonates are favored over carbonates, and adding carbon dioxide to water results in more bicarbonates. Bicarbonate forms of minerals tend to have a less noticeable effect on the taste of water than do carbonates.

Recipes for hard and soft synthetic waters and the median levels of 100 actual US tap waters are shown in Table A-6 (Smith et al., 2002; van der Leeden et al., 1990). However, a challenge arises when multiple salts are dissolved together because of ion effects—one ion affects the saturation of another. The common minerals calcium carbonate and magnesium sulfate are also poorly soluble in water.

Water treatment affects mineral content

Table A-2 summarizes the levels of minerals in large drinking water supplies. In contrast, desalinization using RO treatment, which is designed to remove ions, can result in significantly lower levels of minerals in treated water (Table A-7). Specifically, RO removes calcium and magnesium along with the carbonates, whereas sulfate, chloride, and TDS can still pass through in detectable quantities.

Treatment can have a noticeable effect on tap water taste because the loss of minerals can leave a slick feeling and the loss of alkalinity can leave a drying, bitter taste. Decreased alkalinity also leaves the water more corrosive. Corrosive water, in turn, will increase the leaching of metals from water pipe and plumbing, which can effect the taste of the tap water.

Since desalination of seawater or brackish water removes 95–99% of the minerals and can reduce the pH, it is typically followed by remineralization and restoration of alkalinity. Remineralization consists of lime addition, which contains calcium carbonate and some magnesium carbonate. Caustic soda, bicarbonate, sodium carbonate, phosphates, or silicates can also be used to restore sufficient minerals to the water.

Table A-7   Change in minerals after desalination of seawater by reverse osmosis

Mineral/Ions	Seawater Level mg/L	Desalinated Seawater mg/L
Calcium—Ca^{+2}	434	0.7
Magnesium—Mg^{+2}	1,100	1.7
Sodium—Na^+	10,900	80.8
Silica—SiO_2	40	0.2
Potassium—K^+	251	2.3
Bicarbonate—HCO_3^-	136	1.6
Chloride—Cl^-	19,348	129
Sulfate—SO_4^{-2}	2,150	3.6
Nitrate—NO_3^-	2.4	0.1
Total dissolved solids	34,377	220
pH	7.6	5.7

Source: Desai, 2004

Table A-8   Effect of reverse osmosis on cations and anions in drinking water without a negative impact on taste at room temperature*

Constituent	Philadelphia (Pa.) Annual Average for Drinking Water mg/L	Levels After Treatment 1 mg/L	Levels After Treatment 2 mg/L
Calcium—Ca^{+2}	26	1.5	<1
Magnesium—Mg^{+2}	6	<2	<1
Sodium—Na^+	18	2.5	0.6
Potassium—K^+	1.9	<0.5	<0.5
Chloride—Cl^-	50	<5	<5
Sulfate—SO_4^{-2}	16	<5	<5
Nitrate-N—NO_3^-	1.1	<0.25	<0.25
Total dissolved solids	175	24	4
pH	7.2	6.7	5.7
Alkalinity	41	7	3

*Determined using a screening sensory test by laboratory staff

Deionization, or demineralization, is an ion exchange process used to purify water (Uy & Mesa, 2004). Cationic resins remove divalent cations such as calcium and magnesium and replace them with monovalent ions such as sodium; anionic resins remove anions such as bicarbonate, chloride, and sulfate. Deionized water can have a bitter or flat taste.

Remineralization targets. Bottled waters have gained popularity in the United States (Mackey et al., 2003). Reasons include convenience, perception of healthiness, as well as taste preferences. Bottled water companies typically apply RO as well as ion exchange to purify the water. Thus some purified bottled waters are remineralized using calcium, magnesium, and sodium bicarbonates to make them acceptable for drinking. The levels of minerals found in seven common bottled waters, based on their labels and on laboratory analyses (Whelton et al 2007) are:

- Calcium varied significantly, from <0.1 to 74–97 mg/L and as high as 162–170 mg/L.
- Magnesium varied from <5 to 20–35 mg/L.
- Sodium was usually <10 mg/L.
- Potassium was <5 mg/L.
- Chloride was usually ≤5 mg/L.
- Sulfate was usually ≤15 mg/L, although one water had levels at 38–41 mg/L.
- Nitrate was <1 mg/L.
- TDS ranged from 4–21 up to 210–319 mg/L.
- pH varied from 6.0–6.5 to 7.4–7.5.

More research is necessary

In summary, the presence or absence of any ion, depending on its concentration, can affect the taste of water. The actual concentration of the ion in water, its relative concentration in saliva, and the response of the human senses to the ion are the important factors influencing the water's taste.

Silica, potassium, and nitrate would not be expected to affect the taste of drinking water at the low levels typically found in natural waters and tap waters. Hashimoto et al (1987) suggested that calcium, potassium, and silica have positive effects on the taste of drinking water, whereas sulfate and magnesium have negative effects. De Greef et al (1993) suggested that chloride, carbonates, iron, and magnesium have the greatest negative effects on water taste. Zoeteman (1980) reported that the sulfates, carbonic acid, and bicarbonate have less effect on taste than chlorides and carbonates. It seems that potassium, magnesium, calcium, and sodium in the right balance with bicarbonates would provide good tasting water. The effect of pH, within the typical range for tap water (e.g., 6.5–8.5), primarily controls the speciation of the ionic content.

Table A-8 shows the results of a preliminary study on drinking water in Philadelphia, Pa., where the mineral (cation and anion) content was reduced by RO treatment. In both cases, the taste of the tap water after treatment (at room temperature) was considered good, acceptable, and refreshing. The loss of minerals did not have a negative effect on taste as determined by comparison with bottled water and by interviews with laboratory staff after they consumed the water.

More research is needed with a focus on the natural minerals in drinking water, their effects on water's taste both individually and in combination, individual preferences for water of varying mineral content, and how the purification of water can affect both these minerals and the taste of the water. Such research could produce models to predict the taste of water based on the minerals (cations and anions) present. A proposed schematic representing the ion content of saliva as a reference point for taste is shown in Figure A-1. This is a potential basis for modeling the effects of ions on water taste. For example, an individual's taste rating could change as the concentration of magnesium reaches and exceeds the level found in that individual's saliva. If magnesium plays a minor role in saliva, low levels in water (less than that found in saliva) might have no effect on taste. However, magnesium can readily exceed the levels found in saliva (as it would in mineral waters) and have a negative effect on taste (unless the consumer of the water prefers the mineral taste). Sodium is abundant in saliva and therefore would need to reach levels well above tap water levels to exceed what would be found in saliva and produce a negative effect on taste. Finally, because little potassium is typically found in water, its concentrations may never exceed those found in saliva and therefore would be relatively unimportant to taste. Because people's sensitivities to tastants, and their saliva concentrations of cations and anions can change within a day and across several days for a variety of reasons (e.g., diet and health), the model as shown in Figure A-1 must account for this variability by showing ranges of responses instead of absolute values or threshold lines.

The need to understand the role of minerals in the taste and nutritional value of drinking water is timely for the drinking water industry. The growing uses of RO and water blending and the declining availability of freshwater put a more urgent emphasis on the role of drinking water in the health and quality of life of populations worldwide.

Taste-affecting ions in drinking water

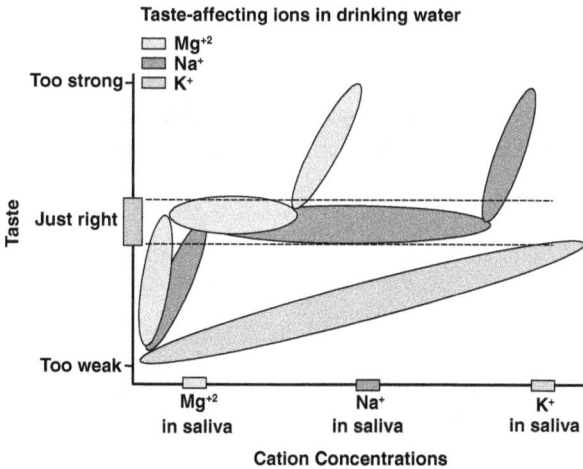

K+—potassium, Mg+2—magnesium, Na+—sodium

Figure A-1 Model showing how cations, in comparison to saliva content, affect the taste of water

Acknowledgment

The authors thank Dr. JaeHee Hong from the Department of Food Science and Technology, Virginia Tech., for her technical insights and suggestions.

About the authors

Gary A. Burlingame *(to whom correspondence should be addressed) manages both the aquatic biology and organic chemistry laboratories as an Administrative Scientist at the Bureau of Laboratory Services, Philadelphia Water Department, 1500 E. Hunting Park Ave., Philadelphia, PA 19124; e-mail gary.burlingame@phila.gov. Burlingame has BS and MS degrees in Environmental Science from Drexel University. An active member of both AWWA and the International Water Association (IWA), he is the current secretary of IWA's specialty group, Off-Flavours in the Aquatic Environment, and chair of the Standard Methods Joint Task Group for Section 2170, Flavor Profile Analysis. Andrea M. Dietrich is a professor at Virginia Tech, and is current chairperson of the AWWA Taste and Odor Committee. Andrew J. Whelton is a consultant and doctoral student at Virginia Tech.*

Footnotes

[1]http://www.dent.ualberta.ca/McGraw/PATH/355/Saliva_Factor.htm

[2]http://www.criminology.fsu.edu/journal/hold.html

[3]http://www.betteroralhealth.info/orbit-complete/saliva.html

References

Argarwal, R.P. & Henkin, R.I., 1987. Metal Binding Characteristics of Human Salivary and Porcine Pancreatic Amylase. *Jour. Biol. Chem.,* 262:6:2568.

AWWA Taste & Odor Committee, 2002. Committee Report: Options for a Taste and Odor Standard. *Jour. AWWA,* 94:6:80.

Brunstrom, J.M. et al., 1997. Mouth-state Dependent Changes in the Judged Pleasantness of Water at Different Temperatures. *Physiol. & Behavior,* 61:5:667.

Bruvold, W.H. & Daniels, J.I., 1990. Standards for Mineral Content in Drinking Water. *Jour. AWWA,* 82:2:59.

Bruvold, W.H. & Pangborn, R.M., 1966. Rated Acceptability of Mineral Taste in Water. *Jour. Appl. Psychol.,* 50:1:22.

Chien, N. & Wan, Z., 1999. *Mechanics of Sediment Transport.* ASCE Press, Reston, Va.

Christensen, C.M. et al., 1987. Salivary Changes in Solution pH: A Source of Individual Differences in Sour Taste Perception. *Physiol. & Behavior,* 40:2:221.

Claes, M. et al., 1997. Inventory of Regulations and Analysis Techniques for Trace Elements in Some Water Supply Companies in Eastern and Western Europe. *Sci. Total Envir.,* 207:141.

Cohen, J.M., et al., 1960. Taste Threshold Concentrations of Metals in Drinking Water. *Jour. AWWA,* 52:5:660.

Conry, B., 2004. Designing Desal for Hotels, Resorts and Coastal Customers. *Water Technol.,* 64:2.

Cuppett, J.D. et al., 2006. Evaluation of Copper Speciation and Water Quality Factors That Affect Aqueous Copper Tasting Response. *Chemical. Senses,* 31:7:689.

DeGreef, E. et al., 1983. Drinking Water Contamination and Taste Assessment by Large Consumer Panels. *Water Sci. & Technol.,* 15:13.

Delwiche, J. & O'Mahony, M., 1996. Changes in Secreted Salivary Sodium Are Sufficient to Alter Salt Taste Sensitivity: Use of Signal Detection Measures With Continuous Monitoring of the Oral Environment. *Physiol. & Behavior,* 59:4:605.

Desai, S., 2004. Desalination: A Technological Overview. *Water Technol.,* 51:12.

Dietrich, A.M. et al., 2004a. *Practical Taste-and-Odor Methods for Routine Operations: Decision Tree.* AwwaRF, Denver.

Dietrich, A.M. et al., 2004b. Health and Aesthetic Impacts of Copper Corrosion on Drinking Water. *Water Sci. & Technol.,* 49:20:55.

Dietrich, A.M., 2006. Aesthetic Issues for Drinking Water. *Jour. Water & Health,* 4: Supplement 1:11.

Doty, R.L. et al., 1984. Smell Identification: Changes With Age. *Sci.*, 226:12:1441.

Fast, K., et al., 2002. New Psychophysical Insights in Evaluating Genetic Variation in Taste. *Olfaction, Taste and Cognition.* Cambridge University Press, UK.

Friel, E.N. & Taylor, A.J., 2001. Effect of Salivary Components on Volatile Partitioning From Solutions. *Jour. Agriculture Food Chemistry,* 49:3898.

Gilbertson, et al., 2006. Water Taste: The Importance of Osmotic Sensing in the Oral Cavity. *Jour. Water & Health,* 4:1:35.

Guinard, J. et al., 1998. Relation Between Parotid Saliva Flow and Composition and the Perception of Gustatory and Trigeminal Stimuli in Foods. *Physiol. & Behavior,* 63:1:109.

Hashimoto, S. et al., 1987. Indices of Drinking Water Concerned With Taste and Health. *Jour. Fermentation Technol.,* 65:2:185.

Hubbard, R.W. et al., 1998. Palatability of Drinking Water: Effects of Voluntary Dehydration. US Army Res. Inst. Environ. Med., Natick, Mass.

Keast, R.S.J. et al., 2003. The Effect of Zinc on Human Taste Perception. *Jour. Food Sci.,* 68:5:1871.

Koseki, M. et al., 2003. Sensory Evaluation of Taste of Alkali-ion Water and Bottled Mineral Waters. *Jour. Food Sci.,* 68:1:354.

Köster, E.P., 2002. The Specific Characteristics of the Sense of Smell. *Olfaction, Taste, and Cognition.* Cambridge University Press, UK.

Lawless, H.T. & Heymann, H., 1998. *Sensory Evaluation of Food.* International Thomson Publishing, New York.

Lawless, H.T. et al., 2003. The Taste of Calcium and Magnesium Salts and Anionic Modifications. *Food Quality & Preference,* 14:319.

Lawless, H.T. et al., 2005. Metallic Taste From Electrical and Chemical Stimulation. *Chem. Senses,* 30:185.

Lim, J. & Lawless, H.T., 2005. Oral Sensations From Iron and Copper Sulfate. *Physiol. & Behavior,* 85:308.

Lockhart, E.E. et al., 1955. The Effect of Water Impurities on the Flavor of Brewed Coffee. *Food Res.,* 20:598.

Mackey, E.D. et al., 2003. Consumer Perceptions of Tap Water, Bottled Water, and Filtration Devices. AwwaRF, Denver.

Matsuo, R., 2000. Role of Saliva in the Maintenance of Taste Sensitivity. *Critical Reviews in Oral Biol. & Medicine,* 11:2:216.

NAS (National Academy of Sciences), 2005. *Dietary Reference Intakes for Water, Potassium, Sodium, Chloride, and Sulfate.* Panel on Dietary Reference Intakes for Electrolytes and Water, Standing Committee on the Scientific Evaluation of Dietary Reference Intakes, Food and Nutrition Board, Institute of Medicine, The National Academies. Natl. Academies Press, Washington.

Olivares, M. et al., 2001. Nausea Threshold in Apparently Healthy Individuals Who Drink Fluids Containing Graded Concentrations of Copper. *Regulatory Toxicol. & Pharmacol.,* 33:271.

Pangborn, R.M. & Bertolero, L.L., 1972. Influence of Temperature on Taste Intensity and Degree of Liking of Drinking Water. *Jour. AWWA,* 64:8:511.

Pedersen, A.M. et al., 2002. Saliva and Gastrointestinal Functions of Taste, Mastication, Swallowing and Digestion. *Oral Disease,* 8:3:117.

Pizarro, F. et al., 1999. Acute Gastrointestinal Effects of Graded Levels of Copper in Drinking Water. *Envir. Health Perspectives,* 107:117.

Preetha, A. & Banerjee, R., 2005. Comparison of Artificial Saliva Substances. *Trends in Biomaterial & Aritficial Organs,* 18:2:178.

Ricter, C.O. & MacLean, A., 1939. Salt Taste Threshold of Humans. *Amer. Jour. Physiol.,* 126:1:1.

Schafer, E. & Zandbiglari, T., 2003. Solubility of Root-canal Sealers in Water and Artificial Saliva. *Intnl. Endodontic Jour.,* 36:660.

Smith, D.V. & Margolskee, R.F., 2001. Making Sense of Taste. *Scientific American,* 284:3:32.

Smith, E.J. et al., 2002. Methods for Preparing Synthetic Freshwaters. *Water Res.* 36:5:1286.

Stumm, W. & Morgan, J.J., 1970. *Aquatic Chemistry: An Introduction Emphasizing Chemical Equilibria in Natural Waters.* John Wiley & Sons Inc., New York.

Szlyk P.C. et al., 1990. Patterns of Human Drinking: Effects of Exercise, Water Temperature, and Food Consumption. *Aviation, Space, & Envir. Medicine,* 61:1:43.

Szlyk P.C. et al., 1989. Effects of Water Temperature and Flavoring on Voluntary Dehydration in Men. *Physiol. & Behavior,* 45:639.

Tenovou, J.O., 1989. *Human Saliva: Clinical Chemistry and Microbiology,* Vol. 1. CRC Press Inc., Boca Raton, Fla.

USEPA (US Environmental Protection Agency), 1979. National Secondary Drinking Water Regulations. EPA-570/9-76-000, Washington.

USEPA, 2003a. Drinking Water Advisory: Consumer Acceptability Advice and Health Effects Analysis on Sodium. EPA-822-R-03-006, Washington.

USEPA, 2003b. Drinking Water Advisory: Consumer Acceptability Advice and Health Effects Analysis on Sulfate. EPA-822-R-03-007, Washington.

Uy, M.C. & Mesa, D.A., 2004. Basics of Deionization. *Water Technol.,* 39:1.

van der Aa, M., 2003. Classification of Mineral Water Types and Comparison With Drinking Water Standards. *Envir. Geol.,* 44:5:554.

van der Leeden, F. et al., 1990 (2nd ed.). *The Water Encyclopedia,* Lewis Publishers Inc., Chelsea, Mich.

van Ruth, S.M. & Roozen, J.P., 2000. Influence of Mastication and Saliva on Aroma Release in a Model Mouth System. *Food Chemistry,* 71:339.

Wiesenthal, K.E. et al., 2007. Characteristics of Salt Taste and Free Chlorine or Chloramine in Drinking Water. *Water Sci. & Technol.,* 55:5:293.

Whelton, A.J. et al., 2007. Minerals in Drinking Water: Impacts on Taste and Importance to Consumer Health. *Water Sci. & Technol.,* 55:5:293.

WHO (World Health Organization), 2004a. Nutrient Minerals in Drinking Water and the Potential Health Consequences of Consumption of Demineralized and Remineralized and Altered Mineral Content Drinking-Water: Consensus of the Meeting. WHO, http://www.who.int/water_sanitation_health/dwq/nutconsensus.

WHO, 2004b (3rd ed.). *Guidelines for Drinking-Water Quality,* Vol. 1. WHO, Geneva, Switzerland.

Yamaguchi, S., 1991. Basic Properties of Umami and Effects on Humans. *Physiol. & Behavior*, 49:833.

Zellner, D.A. et al., 1988. Effect of Temperature and Expectations on Liking for Beverages. *Physiol. & Behavior*, 44:1:61.

Zoeteman, B.C.J., 1980. *Sensory Assessment of Water Quality*. Pergamon Press, New York.

Appendix B

From *Opflow*, December 2010

TDS and Temperature Affect Consumer Taste Preferences

Although treatment and regulations drive the drinking water industry, most consumers judge drinking water based on taste, odor, and visual qualities rather than treatment or reported water quality parameters. It's important, therefore, to understand the qualities of drinking water that produce sensory reactions.

by Conor D. Gallagher and Andrea M. Dietrich

Consumers have high expectations of drinking water, including safety and sanitation, as well as aesthetics and taste. Because consumers desire consistency in their food and beverage products, changes in water quality resulting from mineral content or disinfectant levels are so noticeable that they can result in consumer complaints. Accordingly, a utility's ability to produce an acceptable, consistent product is a big part of gaining and keeping the public's trust. Consumers notice when water doesn't taste right. A recent study at Virginia Tech investigated the influence of total dissolved solids (TDS) and temperature on consumers' ability to discriminate water based on taste and to determine their preferences.

TDS Levels

Minerals, the major contributor to water taste, usually enter natural waters through weathering or erosion of rock and soil but may also come from man-made sources, such as road salt or industrial discharges. The mineral content of drinking water usually isn't altered much from source to tap, but desalination and blending are changing that paradigm. The mineral content of drinking water is frequently measured as TDS, which includes common cations such as calcium, magnesium, potassium, and sodium, as well as anions such as carbonate, bicarbonate, chloride, nitrate, sulfate, and silicates.

VOCABULARY
SENSORY TEST METHODS

Forced Choice. In a sensory difference test, it's good practice to require assessors to choose even when they're uncertain. If they're allowed to respond that
they don't know which sample to choose or that there are no differences among
the samples, the ability to give an opinion becomes confused with being able to
detect the sensory attribute being studied. Also, there is ambiguity about how
"don't know" responses should be used in the analysis. In a forced-choice procedure, assessors are instructed not to opt out of choosing and to guess at random
if necessary.

Triangle Test. The triangle test is a commonly used sensory-difference test. In
this test, three samples are presented to each assessor—two of which are identical and one that's different in some way. Assessors are instructed to indicate
which of the three is odd. If the difference is undetectable to the assessor, the
probability of making the correct choice is one in three (0.333). If correct choices
are made more than one time in three, the best estimate of the probability of making a correct choice is greater than 0.333 and constitutes some evidence that
the difference between the two types of sample is detectable. Other things being
equal, this evidence is more convincing if there's a greater proportion of correct
choices or if the number of trials is greater. If the proportion is correct and the
number of trials is sufficiently great, the result may be statistically significant.

Source: StatBasics (www.statbasics.com/difftest/glossary.htm)

The US Environmental Protection Agency and Health Canada
limit TDS to a 500-mg/L maximum, and the World Health Organization established a TDS maximum of 1,000 mg/L. Increasing
consumer complaints prompted the Taiwan Environmental Protection Agency to reduce the maximum TDS level from 600 mg/L to
250 mg/L.

Acceptable TDS concentrations vary globally and are influenced
by population preferences, from a high of 251–500 mg/L to a low of
less than 100 mg/L. Maintaining TDS at less than 250 mg/L is advisable to avoid a distinct mineral taste. For drinking water in the 100
largest US cities, the range is 22–1,589 mg/L TDS with a median of
186 mg/L. In Canada, TDS levels in drinking water are generally
less than 500 mg/L but may be higher in arid western regions.

Mineral Content and Temperature

Temperature also affects what consumers think about their
water. Tap water's serving temperature generally ranges from 4°C to

about 30°C. North Americans generally have a preference for cold water, but this preference varies globally. Water temperature is known to affect dissolved oxygen content, but the dissolved oxygen content of drinking water has been shown to be unrelated to taste.

Consumers generally detect less of a mineral taste when waters containing 750–1,000 mg/L TDS were chilled to 0°C than at room temperature. A similar temperature-related trend was observed for 11 tap waters ranging from 38 mg/L to 2,460 mg/L TDS, with four of those samples containing less than 500 mg/L TDS and seven samples containing more than 500 mg/L TDS.

Using a forced-choice triangle test (see Sensory Test Methods, page 84) conducted with room-temperature water samples, utility-based consumer panels and sensory experts consistently discerned a difference between treated surface water with ~500 mg/L or higher TDS and desalinated water that was remineralized or blended to much lower TDS values. Reverse osmosis (RO) water from desalination was generally rated as taste free and odor free, with a drying feeling, even when the water was adjusted to up to 100-mg/L alkalinity (as calcium carbonate).

Although major taste factors result from mineral content (TDS) and temperature, the relationship between these factors hasn't been explored much, especially at TDS values greater than 500 mg/L—the upper limit of many secondary standards. In addition, few studies evaluate discriminative *and* preference testing by consumers. This study's objectives were to investigate

- the influence of TDS and temperature on consumers' ability to discriminate among samples.
- consumer preference for waters of different TDS values.
- whether an ability to discriminate relates to preference.

Taste Tests

The study's tasting protocol was approved by the Institutional Review Board at Virginia Tech, and participants granted free and informed consent. Approximately equal numbers of males and females (age range 18–60 years) participated. Subjects tasted two ounces of water in three-ounce coded cups presented in a balanced random fashion. Three waters—A, B, and C—were tested at 4°C and 22–24°C. TDS levels of the waters were 25 mg/L, 36 mg/L, and 500 mg/L for A, B, and C, respectively.

Discrimination Testing. Using the triangle test (see Sensory Test Methods, page 84), subjects were simultaneously presented with three samples of water, consisting of two samples of one water and one sample of another. The samples were coded with random three-letter labels of all possible serving orders (e.g., AAB, ABA, BAA, BAB, BBA, ABB). The subjects were instructed to taste the samples in the order presented and select the one sample that was different.

For the discrimination test, 318 subjects participated, most of whom were North American. Six groups of 53 subjects evaluated one combination of samples at one temperature—Waters A and B, A and C, and B and C. The sample size and significance was based on alpha = 0.05, beta = 0.1, and pd = 0.3; N critical = 25 correct responses out of 53.

Preference Testing. Using the paired-sample test, subjects were presented with two samples of two different waters at 4°C or 22–24°C. The panelists were instructed to taste the samples and choose the one they preferred. The samples were coded with random three-letter labels, and the presentation order was balanced among panelists by using both possible serving orders (AB and BA).

For the preference test, 390 subjects participated. Six groups of 65 subjects evaluated one pair of waters at one temperature: Waters A and B, A and C, and B and C. The sample size and significance was based on alpha = 0.05, beta = 0.1, and pd = 0.3; N critical = 42 correct responses out of 65.

Characteristics of Test Waters. The research used commercially available bottled water that had been refrigerated or stored at room temperature. Bottled water was used for convenience, as it was available in sealed bottles, didn't require dechlorination or refrigeration, and had reproducible water quality. Bottles were stored for up to two weeks before use and weren't opened until it was time to pour the contents into cups for subjects to taste. TDS, organic carbon, inorganic carbon, and pH were measured in a laboratory (Table B-1).

Taste Test Results

Table B-2 presents data for discrimination testing using the triangle test. Table B-3 presents data for preference testing using the paired sample test.

The results demonstrated that—at the 95 percent confidence level—subjects couldn't distinguish the three 4°C waters, nor did they prefer one sample over the other in the paired comparison test. When the three samples were 22–24°C, the subjects couldn't distinguish between the two low-TDS waters (A and B) and didn't have a preference for either low-TDS water.

Table B-1 Water Quality Data. The researchers used commercially available bottled water.

Water Quality Parameter	Water A	Water B	Water C
pH	4.81	4.99	7.28
TOC, mg/L	0.055	0.16	0.05
TDS, mg/L	2.8	31	524
HCO_3, mg/L	0.71	0.22	463

Table B-2 Triangle Test Results. Discrimination was based on TDS and temperature. (N = 53; N critical = min 25 correct responses)

	4°C		22–24°C	
Test Waters	Correct Responses	Able to Discriminate	Correct Responses	Able to Discriminate
A vs. B	24	No	19	No
A vs. C	24	No	26	Yes
B vs. C	23	No	35	Yes

Table B-3 Paired Comparison—Preference Test. Subjects didn't prefer any water even when they could distinguish a taste based on varying TDS levels. (N = 65 for each water pair; N critical = 42 correct responses)

	4°C		22–24°C	
Test Waters	Preference	Ratio*	Preference	Ratio
A vs. B	No	28/37	No	29/36
A vs. C	No	32/34	No	40/25
B vs. C	No	35/31	No	36/29

*Ratio = number of subjects who indicated water was their preference; for example, for A vs. B at 4°C, 28 subjects chose Water A and 37 subjects chose Water B.

In contrast, at 22–24°C, subjects could distinguish between the high-TDS-level water and either low-TDS water (A vs. C or B vs. C), but there was no preference for 22–24°C water. The results for all combinations—including chilled (4°C) and unchilled water (22–24°C)—indicated the subjects didn't prefer any water even when they could distinguish a taste based on varying TDS levels, from low (25–36 mg/L) to moderate (560 mg/L).

Interestingly, certain individuals were confident in distinguishing the waters and their preference for one water over another, illustrating that some consumers had sensitive palates and strong TDS preferences. The water that elicited the most negative responses was high-TDS Water C. However, pH is unlikely to be the reason for the ability to discriminate Water C from Waters A and B, because the main carbonate species in all three waters was bicarbonate, which is much less flavorful than carbonate.

What Does It Mean?

The research indicates that utilities should consider how TDS concentrations affect consumers' perceptions. If a treatment change is planned that will significantly alter the historical TDS concentration (such as implementation of RO, blending practices, or alternative source water), a utility should inform customers of the change and consider performing taste tests.

Authors' Note: This work was supported by the National Science Foundation under Grant No. 0755342 in support of Conor D. Gallagher and Virginia Tech's Institute for Critical Technologies and Applied Science, which provided materials and instrumentation.

From *Opflow*, June 2006

Multiple Barriers for a Smelly Situation

by Mark A. Waer

Public confidence and acceptance of drinking water are largely affected by the aesthetics of the drinking water at the tap. Consumers tend to react more dramatically to taste and odor episodes caused by geosmin and 2-methylisoborneol (MIB) than to odors consistently present in drinking water, such as chlorinous odors. Early detection of potential odor-causing contaminants is essential to avoid problems at consumers' taps.

Although several processes effectively remove taste-and-odor–causing compounds, it is difficult and expensive to achieve a high percentage of removal with a single process. Some utilities have installed preventive treatment processes, such as ozone, to eliminate the occurrence of taste and odor episodes. These processes may be costly to operate, and high concentrations of taste and odor compounds may still get past them.

Installing multiple processes that are integrated in an optimized treatment scheme may be a better option to provide reliable taste and odor control at decreased capital and operating costs. These processes can be optimally configured to produce effective multiple barriers against taste and odor.

In addition to episodes caused by the earthy, musty-smelling geosmin and MIB algae blooms, other taste and odor events may occur, often during runoff. These events may be characterized by smell as swampy, septic, medicinal, grassy, or cucumber-like. Although the compounds that cause these odors are not always known, several are, including two algal metabolites: 2,4-nonadienal, a chemical also formed when cucumbers and similar vegetables are sliced, and *cis*-3-hexenal, the odor produced when grass and other vegetation is cut or damaged. Most other odors can be effectively managed using a single process, such as oxidation with chlorine, ozone, potassium permanganate, or chlorine dioxide, or adsorption using powdered activated carbon or granular activated carbon.

Activated Carbon

The main processes for taste and odor control are adsorption and oxidation. In the adsorption category are powdered activated carbon

(PAC) and granular activated carbon (GAC). The capital costs for PAC can be low, because a simple chemical feed system may be all that is necessary for introducing PAC to the water for adequate contact time. Higher doses of PAC are required to adequately remove geosmin and MIB, however, so the operating costs for PAC may be high.

GAC requires higher capital costs for the construction of contactors and is more costly per pound, but GAC is more effective than PAC and may have lower operating costs for handling persistent tastes and odors such as geosmin and MIB. GAC has limited ability for handling large concentration spikes of taste and odor compounds because it can release these compounds once a spike has passed. Because equilibrium is established between the surface of the GAC and the water, when the concentration in the water is reduced, some of the adsorbed compound may be released from the GAC surface. This is called the chromatographic effect.

One PAC application point for taste and odor control is at the chemically treated water influent to the flocculation tank. By feeding the PAC after coagulant addition, the problems of chemical and background organic interference may be minimized while still providing an acceptable contact time.

A Texas utility using PAC for removing MIB found that about 20 mg/L of PAC will result in about 40 percent MIB removal, while a dosage of 100 mg/L is needed to remove 90 percent of the MIB. Adding this much PAC can raise concerns about handling the PAC and the additional residual solids. Use of GAC would generate fewer handling issues but may introduce new issues concerning GAC replacement.

Oxidation and UV

In the oxidation category, ozone and advanced oxidation processes (AOPs) are the most effective at controlling tastes and odors. At low dosages, ozone can achieve moderate to high removal of taste and odor compounds. To attain the optimal removal necessary to handle high influent concentrations, the ozone dose can be increased or hydrogen peroxide can be fed to convert the ozone to an AOP.

Each of these additions will result in increased costs and may result in a compliance issue with the bromate standard. Under certain conditions ozone or an AOP will react with bromide in the

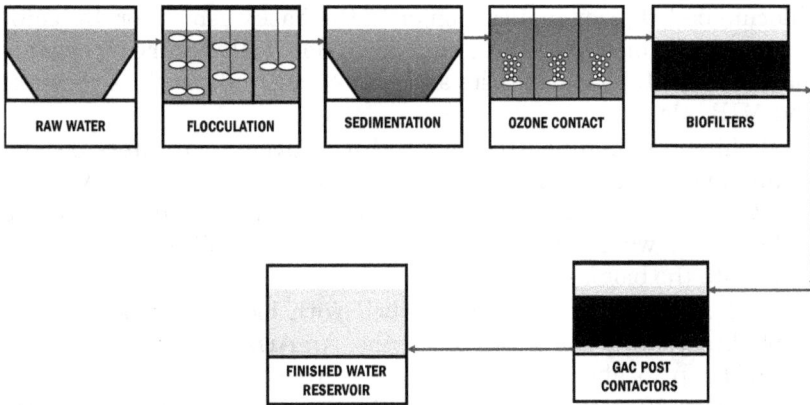

Figure B-1 Process Schematic for Ozone, Biofiltration, and GAC

water to form bromate, which is regulated to an MCL of 10 µg/L by Stage 1 of the Disinfectants and Disinfection Byproducts Rule. Use of ultraviolet (UV) light along with peroxide, another AOP, will not create bromate, but it is costly because of the high intensity of UV light necessary and the additional cost of hydrogen peroxide.

Ozone is optimally incorporated into the treatment process after sedimentation and before filtration. Following ozone, it is common to encourage biofiltration to remove compounds that are easily assimilated by bacteria, thereby lowering the potential for bacterial growth in the distribution system. An Arizona utility oxidizing MIB with ozone achieved 45 percent removal with a low ozone dosage (about 1 mg/L), but much higher dosages were needed to achieve complete removal, and dosages at high levels can generate unacceptable levels of bromate.

The UV–peroxide oxidation process occurs just before the finished water is released to the distribution system, after conventional filtration, with the peroxide added just before the UV treatment. The process effectively oxidizes MIB at a constant UV dose. Low peroxide dosages (about 3 mg/L) can achieve 60 percent MIB oxidation, but dosages of 8 mg/L may be needed to accomplish 90 percent removal.

Biofiltration

Biofiltration is essentially a process that uses GAC as a media to support biological growth. For this to work, it is important that no chlorine be fed prior to filtration or in the backwash water. In biofiltration, while biomass is breaking down naturally occurring organic matter, co-metabolism of taste and odor compounds, such as geosmin and MIB, also occurs.

Biofiltration can be accomplished in conventional filters on any media, such as sand or anthracite, although some studies indicate that GAC creates a more robust biomass. Chlorine for disinfection is added to the water after it passes through the biofilters so as not to inactivate the biomass.

MIB removal can be accomplished with biofiltration at varying empty-bed contact times, or bed depths. An Arizona utility increased removal (up to about 67 percent) by increasing empty-bed contact times up to 10 min, but greater contact times did not significantly aid removal.

For MIB concentrations requiring less than 60 percent removal, biofiltration provides the removal at no additional operating expense. If higher removals are needed, PAC at 10–20 mg/L can be added for extra removal. And at the highest MIB concentration, UV–peroxide oxidation would be implemented. The cost to achieve 90 percent MIB removal would be less than either PAC or UV–peroxide oxidation alone, and the cost for lower removals would be low.

Combining Processes

Combining two or three of these processes can prove less costly and more effective at achieving high removals than using only one process. For example, adding 20 mg/L PAC prior to flocculation (55 percent removal), replacing conventional filtration with biofiltration (65 percent removal), and including UV–peroxide oxidation at peroxide dosages of 3 mg/L (65 percent removal) at the end of the treatment process will result in nearly 95 percent removal at a low cost.

Another option is combining ozone, biofiltration, and GAC postfiltration contactors for control of tastes and odors. In this case, reliability is the main goal, with costs a secondary concern. Ozone and biofiltration provide an excellent and low-cost solution for removing high percentages of taste and odor compounds, while the GAC serves as a backstop against any taste and odor escaping the plant. Although GAC is normally expensive to install and operate, in this case the operational cost is minimized by maintaining a long replacement interval through the use of upstream processes.

In conclusion, although several separate processes may be used to effectively remove taste and odor compounds, attaining high-percentage removal of problematic compounds with a single process is difficult and expensive. Combining various processes can achieve reliable taste and odor removal at reduced operating costs.

For More Information

The AWWA Bookstore has several items that relate to taste and odor problems, including these:

- *Taste and Odor: An Operator's Toolbox*, video
- *Water Quality Complaint Investigators Field Guide*
- *Practical Taste and Odor Methods for Routine Operations: Decision Tree*, AwwaRF report
- M7, *Problem Organisms in Water: Identification and Treatment*
- M12, *Simplified Procedures for Water Examination*

From *Opflow*, October 2003

What the Nose Knows: Strategies for Taste-and-Odor Testing Methods

by Andrea M. Dietrich, Gary A. Burlingame, and Robert C. Hoehn

As more consumers express their demand for drinking water that is both safe and aesthetically pleasing, testing provides a means to monitor both water quality and water security. Sensory monitoring of the water quality in the distribution system is important for potentially detecting off-tastes and odors, including taste compounds leaching from corroding metals or taste-and-odor compounds from plastic pipes or microbial contamination.

Sensory methods with detailed standard operating procedures have existed for many years. Previous editions of *Standard Methods for the Examination of Water and Wastewater,* as well as the current edition (1998), include the threshold odor number test, flavor profile analysis, flavor threshold test, and flavor rating assessment. An AWWA Research Foundation study, *Practical Taste-and-Odor Methods for Routine Operations: Decision Tree,* to be released next year, has developed new sensory methods for the water industry that includes the attribute rating test, 2-of-5 test, rating method for distribution system odors in comparison to a control, and triangle test. Many methods were derived from programs implemented by the food and beverage industries. The sensory methods can successfully be applied to solving many water industry problems, including

- Routine monitoring of raw or finished water to determine quality and detect aesthetic changes
- Evaluating treatments designed to remove tastes or odors
- Tracking taste-and-odor problems in watersheds
- Evaluating water samples from customers' premises
- Early warning of the occurrence of taste and odor in raw water
- Documenting the intensity of individual odorants during odor events

Just as no single analytical chemistry method can detect all chemical contaminants in water under any given set of conditions,

Table B-4 Matching Objectives to the Test

Objectives	Potential Tests to Use
Bench-scale studies on the approval of materials for leaching and drinking water contact	AWWARF, FPA
Approval of the release of water in new mains or new and rehabilitated facilities	FPA, FRA, 2-of-5, RMD
Screening source water for problems	FPA, TON, ART
Screening drinking water for problems	FPA, FRA, 2-of-5, RMD, ART
Customer complaint sample evaluation	FPA, FRA, 2-of-5, RMD
Odor or taste threshold determination	FPA, FTT, ASTM, △Test
Correlation to analytical test results	FPA, FRA, 2-of-5, RMD
Developing a baseline of the sensory characteristics of a water	FPA
Taste-and-odor event response	FPA, FRA, 2-of-5, RMD, ART
Evaluation of treatment effectiveness bench-scale or full-scale	FPA, FRA, 2-of-5, ART
Routine drinking water quality control	FPA, FRA, 2-of-5, RMD, △Test
Evaluating analysts' sensory abilities	△Test, S&S, CBS

FPA:	Flavor Profile Analysis, as described in *Standard Methods* 2170
TON:	Threshold Odor Number method, as described in *Standard Methods* 2150
FTT:	Flavor Threshold Test, as described in *Standard Methods* 2160
FRA:	Flavor Rating Assessment, as described in *Standard Methods* 2160
△Test:	Triangle Test
S&S:	Sensonics Scratch and Sniff Test (P.O. Box 112 Haddon Heights, NJ 08035)
CBS:	Carolina Biological Supply Odor Test Kit (2700 York Rd., Burlington, NC 27215)
2-of-5:	2-out-of-5 Test, as described in AwwaRF, 2004
ART:	Attribute Rating Test, as described in AwwaRF, 2004
RMD:	Rating Method for Distribution System Odors in Comparison to a Control, as described in AwwaRF, 2004
ASTM:	Threshold Method E679-91, as described in *Standard Practice for Determination of Odor and Taste Thresholds by a Forced-Choice Ascending Concentration Series Method of Limits*, ASTM 1992.
AwwaRF:	Development of a Method for Taste-and-Odor Materials Evaluation AwwaRF 2002b

These abbreviations are used in Tables B–4.

no single sensory method can provide all answers to taste-and-odor questions. The water industry needs a robust toolbox of methods to evaluate sensory properties of water and from which to choose the most appropriate methods to meet utility needs. Several simple, yet reliable, taste-and-odor evaluation methods are currently available to water-treatment plant personnel for day-to-day monitoring.

Sensory Methods

The various sensory methods used in the water industry are listed below in alphabetical order, along with the recommended temperature at which each test should be performed. Maintaining the proper temperature is critical to obtaining reproducible results during taste-and-odor evaluations. Also, chlorine in water samples may interfere with the perception of other odors, so dechlorination prior to the taste-and-odor analysis is recommended when the focus is evaluating odors other than chlorine.

American Society for Testing and Materials, Threshold Testing, ASTM *(ASTM E679-91)*: No single standardized method or sensitivity test for determining thresholds of odorants in drinking water is recognized by the drinking water community. The peer-reviewed method ASTM E679-91, Standard Practice for Determination of Odor and Taste Thresholds by a Forced-Choice Ascending Concentration Series Method of Limits (ASTM 1992), is a rapid test for estimating sensory thresholds and addresses both detection and recognition thresholds. The method uses a series of samples from low to high concentrations selected to bracket the expected threshold level. Each panelist performs a triangle test (one sample with the odor in it and two without) at each concentration. From the data of correct responses, an individual's odor threshold is estimated, and a group estimate is calculated using the geometric mean of individual thresholds.

Attribute Rating Test, ART *(AwwaRF 2004)*: ART is a paired-comparison test by which one identifies specific odorants (e.g., geosmin or 2-MIB) in water samples warmed to 45°C and rates the odor intensity as less than, equal to, or greater than that of standards containing a known concentration of the odorant. For geosmin or 2-MIB, 15 ng/L is used because it is the concentration at which many consumers first begin to register complaints. ART is an efficient, focused, cost-effective method that treatment plant personnel can use to monitor raw and treated water for specific odorants and to maintain the finished water's odor below a target concentration.

Carolina Biological Supply Odor Test Kit, CBS: The CBS Sense of Smell Kit can be used to evaluate olfactory sensation and odor discrimination. It consists of 12 reusable odor vials, an instruction manual, and 30 instruction sheets. In 15 to 30 minutes, a small group can test its ability to recognize and identify a variety of familiar odors, study the phenomenon of sensory adaptation, and learn the difference between olfactory and trigeminal sensations. The kit

Table B-5 Decision-making Summary: Sensory Test Resource and Skill Level Requirements

Test Method / Factor	FPA	TON	FTT	FRA	Δ Test	S&S	CBS	2-of-5	ART	RMD	ASTM	AwwaRF
Initial Lab Setup Costs	E	S	S	E	M	M	M	S	S	S	S	S
Materials Requirements	E	S	S	E	S	M	M	S	S	S	S	S
Lab Requirements	E	M	M	S	M	M	M	M	M	M	E	S
Routine Operating Costs	E	S	S	E	M	NA	NA	S	S	S	NA	NA
Time to Set up and Perform a Test	E	S	S	E	E	M	M	S	S	S	E	E
Analyst Participation Requirements	E	M	M	S	S	S	S	M	M	M	E	S
Level of Training	E	S	S	E	S	M	S	M	M	M	E	S
Data Management Requirements	E	S	S	E	E	M	M	S	M	M	E	E
Data Interpretation Requirements	E	S	S	E	S	M	S	S	M	M	E	E
Quality Assurance	E	S	S	S	S	M	M	S	S	S	E	E

M = minimal; no major investment needed
S = significant; for example, for lab setup, equipment such as water bath necessary
E = extensive investment of time or resources
NA = not applicable

can be used for training as well as for evaluating a person's ability to detect and describe common odors.

Flavor Profile Analysis, FPA (Standard Method 2170; AWWA 1993): FPA was first developed for the food industry in the 1950s and standardized as a method for drinking water flavor analysis in *Standard Methods* in 1995. FPA uses a trained panel of four to seven individuals to characterize and assign intensities to tastes and odors. The method requires intensive training and standardization in order to achieve reliable and reproducible results. Samples are evaluated either at room temperature (25°C) or after being warmed to 45°C. Each assessment provides one or more descriptors and an intensity rating for each on a scale of 0 to 12, with 12 being the strongest. After evaluating the sample, panelists discuss their individual opinions.

If at least 50 percent agree on a descriptor, an average for its intensity is calculated. An FPA panel can often identify more than one taste or odor component in a water sample, thus generating a "profile" of the sample's flavor character.

A variation of FPA is Quantitative Descriptive Analysis (QDA). It is similar to FPA, but instead of consensus, panelists provide individual ratings to which statistical analyses are applied. Many utilities train using FPA, but process their data in the manner of QDA.

Flavor Rating Assessment, FRA *(Standard Method 2160)*: FRA estimates acceptability for daily consumption, using a 9-point affective scale for either 4 to 8 trained panelists or consumers. FRA requires a complete preliminary test using known samples to develop coefficients of correlation for each assessor in the trained panel. The method determines the acceptable level of odorants or tastants in water at room temperature. The water is rated by averaging the FRA ratings of each panelist. The action tendency scale ranges from 1 ("I would be very happy to accept this water as my everyday drinking water") to 9 ("I can't stand this water in my mouth and I could never drink it").

Table B-6 Example for Customer Complaint Sample Evaluation: Comparison of Resource Requirements and Skill Requirements for Available Sensory Tests

Test Method Factor	FPA	FRA	2-of-5	RMD
Initial Lab Setup Costs	E	E	S	S
Materials Requirements	E	E	S	S
Lab Facility Requirements	E	S	M	M
Routine Operating Costs	E	E	S	S
Time to Set up and Perform a Test	E	E	S	S
Analyst Participation Requirements	E	S	M	M
Level of Training	E	E	M	M
Data Management Requirements	E	E	S	M
Data Interpretation Requirements	E	E	S	M
Use of Quality Assurance	E	S	S	S

M = minimal; no major investment needed
S = significant; for example, for lab setup, equipment such as water bath necessary
E = extensive investment of time or resources
NA = not applicable

Flavor Threshold Test, FTT (Standard Method 2160): FTT is the application of the ASTM method of limits to drinking water flavor or odor. It estimates the sensory thresholds of detection for odorants in water using only one ascending series of presentations that increase in concentration of the odorant. FTT recommends screening assessors based on sensitivity to n-butyl alcohol and o-chlorophenol, both of which have well-established thresholds. An average threshold is calculated at the first point of detection from many assessors.

Rating Method for Evaluating Distribution-System Odors in Comparison to a Control, RMD (AwwaRF 2004): This is a "difference method," with samples warmed to 45°C, that compares various unknown distribution-sample odors to those odors of a control, which can be any water the utility considers to represent its "ideal" or typical water (e.g., finished water collected at the point of entry to the distribution system). The test is an excellent quality-control test, as it allows panelists to detect very small differences between the odors of samples and the control. If the odors are the same, the water that customers receive has not been changed during transmission. Differences in the odors signal the development of problems in the distribution system. The analyst is required to characterize sample odors using common terminology for describing off-odors.

Sensonics Scratch and Sniff Test, S&S: Three common odors are used in this system, which is an economical way to screen a large number of persons for their odor-detecting abilities. The test is useful for screening out people who have smell deficiencies.

Threshold Odor Number, TON (Standard Method 2150): The most widely used sensory test in the water industry for about 50 years, the TON is the basis of the US Environmental Protection Agency secondary maximum contaminant level for the odor of water (not to exceed a TON of 3). The current procedure in *Standard Methods* (1998) uses odor-free water to dilute the water sample to be tested; the odor is evaluated at either 60°C or 40°C. The total volume (odor-free water volume + sample volume) to be used in the test is 200 mL, and the TON is the number obtained by dividing the total volume by the volume of the odorous sample being evaluated. Thus, if the first dilution in which the odor could be detected contained 5 mL of the sample, the TON would be (195 mL + 5 mL) divided by 5 mL, or TON = 40. The TON is not a measure of odor intensity, but a measure of an odor compound's persistence upon dilution.

Triangle Test, △Test (AwwaRF 2004): The △Test is the most widely used of the forced-choice methods for sensory analysis. This simple, focused test requires minimal training and uses two identical samples and one different sample. Panelists discern the unique sample among the three. For example, a panelist presented with two identical plant effluent waters and one treated water sample from the distribution system is asked to select the sample that is different. The △Test can also be used to screen an individual's ability to detect specific odorants.

2-of-5 Odor Test, 2-of-5 *(AWWARF 2004)***:** This is another forced-choice method that allows an analyst to determine if the odors of two water samples are the same or different. Five flasks are used, two containing the test sample and the other three containing the control water. The sample and control can vary depending on the goals, but the principle is the same, namely that analysts are seeking to determine if the sample odor differs from that of the control. The five flasks are warmed to 45°C and randomly mixed, then placed on a revolving plate and twirled. The analyst then sniffs the samples and sorts them, based on their odor characteristics, into one group of two and another group of three. If the samples are correctly sorted, the two waters are considered to be noticeably different. Finally, the analyst describes the difference by selecting a term from a list of common odor attributes.

Strategies for Selecting the Best Method

Table B-4 (page 95) summarizes typical objectives for using sensory methods and provides options for which tests would be appropriate for use in different situations. For most applications, more than one sensory test method can be used. The choice of the method depends on the time, resources, and personnel that a utility can invest.

Table B-5 (page 96) compares individual sensory tests to the amount of resources that each method requires. The categories are subjective, but they distinguish between minimal, significant, and extensive contributions to make the method work. For example, in the materials requirement, "minimal" indicates the need for typical glassware; "significant" would indicate the need for a specialized piece of equipment such as a water bath; and "extensive" would signify that more dedicated or specialty equipment than a water bath is required.

For a specific need in sensory testing, combine Tables B-4 and B-5 to select the specific application. The requirements of all possible tests can be compared, and the utilities can match their resources to the test requirements to make the most appropriate choice.

An example of this approach is presented in Table B-6 for customer complaint evaluations.

Conclusions

Water utilities should incorporate one or several of the taste-and-odor evaluation methods into their routine water-quality monitoring program. The methods complement more expensive and time-consuming evaluations for chemical and biological monitoring. The most attractive features of sensory methods are their simplicity and potential applications to a wide variety of situations, such as detecting subtle changes in raw and treated water odors, evaluating the effectiveness of treatment processes, and tracking odors in the distribution system. Adoption of these methods may require funding for staff education, odor ability screening, and training through hands-on workshops given at water industry conferences or at individual utilities.

In selected situations, taste-and-odor evaluation methods can be "screening tests" to determine if there is a problem. The 2-of-5 test, △Test, and rating method for evaluating distribution-system odors in comparison to a control are well designed for use as screening tests. For example, if a customer complains of an off-odor in the tap water, a utility could compare the customer sample to the plant effluent with any of the above-mentioned tests as a first step to determine if a difference exists. If none are found, which is often the case, then the odor of the customer's water is like that of the water leaving the plant. If a problem is found, the utility can perform additional testing to determine the source of the difference and to identify the problem odors.

Other sources of valuable information concerning tastes and odors in drinking water include the updated taste-and-odor wheel that succinctly summarizes the current knowledge of the compounds that cause the various odors, testing procedures for evaluating the sensory properties of materials used in plumbing and distribution systems, and numerous publications on identification and treatment.

Acknowledgments

This work was funded by AwwaRF; Djanette Khiari was the project officer. The authors acknowledge the assistance of Thomas Gittelman for reviewing T&O methods used by the water industry. Mentioning of trade names does not imply an endorsement of the product.

Index

Note: *f.* indicates a figure; *t.* indicates a table.